The Stoic Warrior

45 Maxims for Life

FROM

MARCUS AURELIUS

Meditations

Andrii Datsenko

Copyright © 2023 by Andrii Datsenko
All rights reserved.

No portion of this book may be reproduced in any form without written permission from the publisher or author, except as permitted by U.S. copyright law.

Book Cover by Andrii Datsenko
Illustrations are in the public domain

Contents

9   *Preface*

10   *Note on translation*

13   **Maxim 3.13** *(Introduction)*
"Always have your principles at hand"

17   **Maxim 2.1**
"Say to yourself in the early morning: Today I shall encounter a busybody, an ingrate, a bully, a liar, a schemer, a self-seeker"

21   **Maxim 2.4**
"Recognize that limit of your time is set"

25   **Maxim 2.5**
"Resolve to do the work in hand with scrupulous and unaffected dignity"

29   **Maxim 2.11**
"Act like a man who could take leave of life at once"

31   **Maxim 3.4**
"Do not waste what is left of life worrying about others"

35   **Maxim 3.5**
"Don't overdress your thoughts in finest language"

37   **Maxim 4.12**
"Be set to change your course, if one be near to set you right"

39   **Maxim 4.24**
"Do few things — if you seek cheerful calm"

43   **Maxim 4.36**
"Ever behold as all things come to be through change"

47   **Maxim 4.38**
"See clearly what governs mind of wise"

53   **Maxim 5.23**
"Think often of the swiftness with which the things sweep past and disappear"

57 Maxim 6.2
    "Just do the right"

59 Maxim 6.11
    "Don't break the rhythm more than you must"

63 Maxim 6.13
    "Strip the facts bare"

65 Maxim 6.31
    "Sober yourself: recall your senses"

69 Maxim 6.38
    "Meditate upon interconnection of all things"

73 Maxim 6.50
    "Act with a reservation"

77 Maxim 6.53
    "Connect to speaker's words"

79 Maxim 7.1
    "Remember that you've seen it all"

83 Maxim 7.27
    "Reflect upon the blessings you possess"

87 Maxim 7.29
    "Stop acting like a puppet"

93 Maxim 7.38
    "Fret not at things external. For they care nothing"

95 Maxim 7.47
    "Watch the stars in their courses"

99 Maxim 7.48
    "Take a bird's-eye view of earthly things"

105 Maxim 7.56
    "As one who's dead, resolve to live the rest in sweet accord with nature"

109 Maxim 7.57
    "Love what happens only, woven in the web"

117 Maxim 7.59
    "Dig within. Within you is the fount of Good"

121 Maxim 8.2
    "At any action ask, "What does it hold for me?"

127	Maxim 8.32
	"Arrange life act by act"
131	Maxim 8.36
	"Let not the visions from the entirety of life dismay you"
135	Maxim 8.49
	"Tell yourself no more than what's declared by the first impressions"
138	Maxim 8.50
	"Is the cucumber bitter? Toss it out! Complain not"
145	Maxim 9.40
	"Turn your prayers, see what comes."
149	Maxim 10.8
	"Assume your titles"
152	Maxim 10.16
	"Talk not of being good. Be one"
157	Maxim 10.29
	"At each single act pause and ask, if loss of this makes death a terror?"
159	Maxim 10.30
	"Dwell on how you fail too"
163	Maxim 11.18
	"Keep in mind, not being angry, but being kind is manly"
166	Maxim 12.3
	"Learn to live in what alone is life—the present"
171	Maxim 12.6
	"Practice, even when despair of success"
175	Maxim 12.17
	"If not right, do it not; if not true, say it not"
177	Maxim 12.20
	"Do nothing but leading to some social end"
181	Maxim 12.1 *"Fear never beginning to live"*, Maxim 4.17 *"While you live, while you may, become good"*
188	*Special Bonus*
189	*Acknowledgements*
191	*List of Illustrations*

Portret van Marcus Aurelius, John Faber (I), after Peter Paul Rubens, after anonymous, 1691 - 1721

Preface

These are the *Rules for life* of a man who lived in a world that was not so different from our own. A man who faced the same challenges, the same struggles, the same doubts that we face every day. But unlike many of us, this man had a way of looking at the world that allowed him to navigate these challenges with clarity and purpose, to be a true Warrior of the spirit and the mind, for whom the only reason to study philosophy was to put it into practice.

This man was Marcus Aurelius, a Stoic philosopher and Roman Emperor who lived in the second century AD. His book, Meditations, is a collection of personal reflections that he wrote to himself, meant to serve as a guide for how to live a good life.

Marcus Aurelius had an uncanny ability to distill complex ideas into simple, memorable phrases. Within those meditations, there are nuggets of gold—Stoic maxims, short and pithy sayings that encapsulate the wisdom of the Stoics in a concise and easily accessible form.

These maxims are a treasure trove of valuable insights into how to navigate the ups and downs of life and how to achieve a state of inner peace and tranquility. They teach us to focus on what we can control and let go of what we cannot. They emphasize the importance of reason, self-discipline, and virtue and teach us to accept the natural order of things, even when it is not what we would choose for ourselves.

In a world full of distractions and noise, it is more important than ever to have these maxims at our fingertips. This book explores some major Stoic maxims from Marcus Aurelius' Meditations and how they can be applied to our own journey.

Each maxim is presented in a way that is interconnected, so that each maxim builds upon the others. I did my best to approach each Maxim with a sense of imagination and wonder, each one is illustrated with allegorical frescoes of The Northern Renaissance artists to inspire you to think deeply and act wisely. It is my hope that this book will serve as a guide and inspiration for all those who seek to live a happy and smoothly flowing life.

May these Maxims be a light in the darkness, a compass in uncertain times, and a reminder of what truly matters in this world.

Yours truly,
Andrii

Note on Translation

I've retranslated original Greek passages of *Meditations* on my own, comparing and contrasting existing translations to achieve a more authentic rendition.

Maxim

3.13

(Introduction)

"Doctors keep instruments and knives at hand for sudden calls upon their art. ALWAYS HAVE YOUR PRINCIPLES AT HAND FOR THINGS DIVINE AND HUMAN *too."*

Marcus Aurelius. Meditations 3.13

Life is warfare, a wandering, away from home; and after-fame, oblivion. What is to guide us then? One thing, philosophy;
Meditations, 2.17

Stoicism is a way of seeing things. Like any other philosophy, it is a framing put upon reality. But it is not a block on the reality of our emotions; it's not about being "stoical." It's being strong and wise enough to frame things with poise and love.

Why did Marcus Aurelius frame his core beliefs in such beautifully succinct statements? He knew that,

"Whatever you imagine often, shapes your mind: your soul takes the dye of your impressions."
Meditations, 5.16

Or, as it was wrapped by Buddha Gautama, *"What you think, you become."*

Marcus repeats this thought elsewhere, he oftentimes repeats himself. He dyes his mind with precious precepts. It's a simple psychology: to dwell upon an image is to ground the impression in the mind. Your self is stained by what you frequently imagine.

"And almost thence my nature is subdued
To what it works in, like the dyer's hand"
William Shakespeare, Sonnet 111

Stoics are to be vigilant against unreasoned thoughts popping up in their heads. Always ready to replace imperfect judgment with a reasoned one. This is where the maxims step in.

The Stoics craft their beliefs with care, each one a gem, polished and clear, so they stick in the mind and be always accessible—the only way they can be wielded in a world that shifts without warning. Short, sharp statements are like blades that slice through doubt and uncertainty. They are the mental "weapons" against unreasonable thoughts and judgments.

"In applying principles to action, be a fist-fighter, not a swordsman. The swordsman lays by the blade and takes it up again; the fighter's hand is always there, he needs but clench it."
Meditations, 12.9

The stoic maxims are meant to be internalized, are to become an inalienable part of the stoic self, not taken up and set aside at will.

Just like a boxer with his fists, *keep your "weapons" ready at hand.*

Ultimately, they will decide your life—whether an unfulfilling or a happy and smoothly flowing one.

Armed with these "weapons",
your fate is cast,
A life of woe, or joy that lasts.

Only a sick body is cared for, Philips Galle, after Philips Galle, 1610 - 1676
A donkey lies on its sickbed and vomits. He symbolizes the sick human body (Caro) and is nursed by the personification of the lustful person (Homo Carnalis). A doctor is called to the patient and looks at a jar of urine. In the background, the personification of the human neglected soul (Anima) dies in a barn. A devil flies above her and will escort her to hell. Print from a series of four prints about man's care for the body and ignoring the soul.

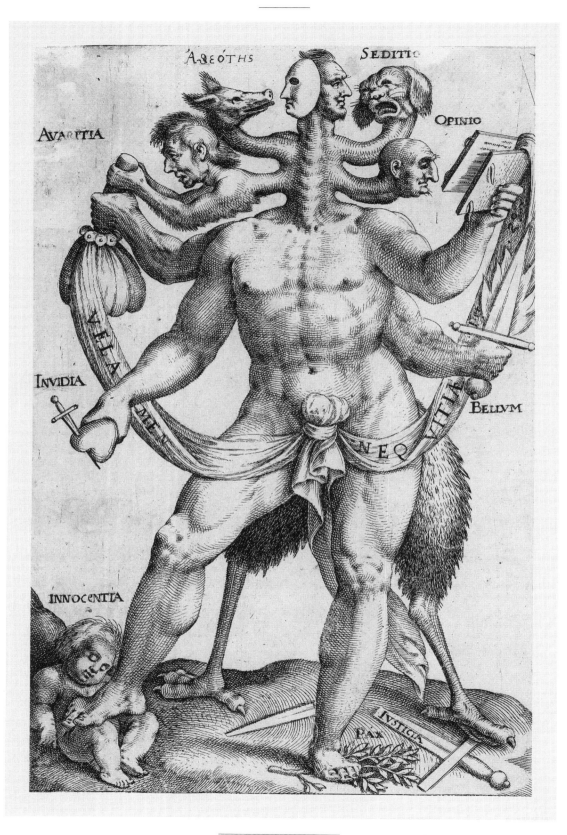

Maxim 2.1

"Say to yourself in the early morning: Today I shall encounter a busybody, an ingrate, a bully, a liar, a schemer, a self-seeker.

All these things have come upon them because they can't tell good from evil."

Marcus Aurelius. Meditations 2.1

The person you see in the mirror each morning may be the most challenging you'll ever meet. But there are more out there, waiting and ready to test you with their demands.

Are you ready to meet them all with poise, grace, and an open heart?

If you take a moment to pause in the morning, to gather your thoughts and your breath, to get ready in heart and in mind, you'll have a better chance of dancing through the day's demanding exchanges with poise and joy.

In fact, that is a special case of *"negative visualization,"* the Stoic exercise of imagining things going awry. The Stoics had a term for this: *Premeditatio Malorum*, or *"the premeditation of evil."*

Allegorical five-headed monster, anonymous, 1575 - 1618
Allegorical five-headed and four-legged monster. This monstrosity has the heads of Avarice (Avaritia), the masked Deceit (Fraus), Rebellion (Seditio) and Delusion (Opinio). The pig's head is indicated by a Greek term (Stupidity / Stupiditas?). In his hands he carries attributes of Envy (Invidia) and War (Bellum). Under his feet he tramples Innocence (Innocentia) and Peace (Pax) and Justice (Justitia)

The idea is to prepare your mind to imagine the worst so you can ultimately see the good and opportunity in all situations, no matter how grim. This will allow you to stay true to your values and your best self instead of panicking and going crazy.

Negative visualization is simply thinking of the worst that could happen.

Think of yourself as dead. You have lived your life *(Maxim 7.56, "As one who's dead, resolve to live the rest in sweet accord with nature")*.

It may sound *frightful* at first. How could negative thoughts—thinking about the worst possible outcomes—make you happier?

Well, the advice is not to dwell on bad things. That would be dreadful. You'd soon become a gloomy mess, complaining about everything.

Negative visualization is not negative thinking.

Premeditatio Malorum assumes that you see into how bad things could get. Then you imagine how you'd handle those low moments and choose the best route to take.

Your options are simple: either you can act on it or you can't. If you can, then imagine the outcome you desire and the obstacles in the way so they won't surprise you when they come. Now they are nothing new to you.

It's like shadowboxing.

The boxer steps into the ring alone and starts visualization drills—he throws punches in the air with a purpose, visualizes the opponent, his moves, and where his fists gonna land.

What's the best response? Counter-jab? Left hook?

Some things, alas, are beyond your control. Negative visualization also helps in hopeless cases. You must learn to accept them with grace, and *"to love what happens only, woven in the web. For no fits better, ever" (Maxim 7.57)*.

In those cases, you could practice something most people often fail to do—learn to manage your expectations *(Maxim 8.50, "Is the cucumber bitter? Toss it out! Complain not")*. You acknowledge that the worst could happen and that you'll be fine. What else can you do?

Now, back to morning rituals: before you go on your journey, ask yourself two sets of questions:

1. *"What if?"*

What could possibly go wrong? What obstacles might come?

These shouldn't be mere rhetorical questions. Imagine in vivid detail what could go wrong and how bad it may get. Explore the darkest corners, the bleakest possibilities of what could go awry. The goal is to be realistic about what could come.

Recognize the worst case as a real option. Imagine surviving it all and being fine with it. Still breathing, still alive, still having a chance at a fulfilling life. See the beauty, the moments of pure delight. The love that surrounds you, the things that make it worth the fight. That will help you stay calm and make wise choices.

The next set is:

2. "What then?"

Now, if the worst *does* happen, what then? How do you fix it?

The goal here is to prepare for the worst. What is the best response?

Devise your plan *(Maxim 8.32, "Arrange life act by act")*. List all scenarios in which you can improve the situation. Never forget: *Fortune falls hard on those who do not expect. The one always alert can easily endure.*

Confidence comes from standing tall and facing facts, not running from the truth, however harsh. External adversity can't truly be bad, the Stoics say, for it's beyond your control. But your reaction—that you can shape, make good, or bad. You train for this—to meet the challenge, to respond with virtue and grace in soul and heart.

Life, after all, is not all joy and success all the time *(Maxim 4.36, "Ever behold as all things come to be through change")*. Things will go wrong someday. Remember Murphy's:

Anything that can go wrong will go wrong.

The only way to cushion the blow is to be ready, make a plan, and stick to it.

It's best to plan for the best and prepare for the worst, even if the bad things never happen. Because you're the only thing you can control.

"Let not the future trouble you. You'll come to it, if so be, armed with the same very reason you now apply to the present."
Meditations, 7.8

P.S. **Marcus goes on in Meditations 2.1:**

"But I considered the nature of the good, and found it beautiful; I beheld the nature of the bad, and I found it ugly. I know that these wrong-doers are by nature my brothers, not by the same blood or seed, but by sharing in the same reason and the spark of divinity.

And so I cannot be injured by any of these. No one can involve me in wrong. Nor can I be angry with my relative, or hate him. For we have come into being for co-operation, as have the feet, the hands, the eyelids, the rows of teeth, upper and lower. It is against nature therefore to oppose each other; and what else is aversion and anger?"

Be kind to those who aren't. They need it most.

Treat every one you meet since now like they'll be dead by dawn.
Be as compassionate and understanding as you can.
Your life will change.

Maxim
2.4

"To recognize that limit of your time is set,

which, if you don't dispel the mists within, will pass away, and you along with it; and chance will never more return."

Marcus Aurelius. Meditations 2.4

The clock ticks on, slow and cruel,
As we count down the hours, feeling like nulls.
We beg and plead for time to fly,
But it only crawls, as we sit and sigh.
Only two more hours, we moan and groan,
Trapped in this cubicle, all alone.
But soon, oh soon, the shift will end,
And freedom, sweet freedom, we will rend.
Until then, we wait and we bide,
Hoping and praying for the day to slide.

"Only two (three, etc.) hours to go!", one of the popular phrases heard in office spaces. People are dying for their time to pass, their shift to be over.

The problem is, whenever someone says something along those lines, they are willing to wish away time they will never earn back.

Memento mori, Raphaël Sadeler (I), after Christoph Schwarz, 1687 - 1749
A putto sits between an hourglass and a skull in front of a tomb with a smoke pot and a flower vase.

Time is more than just a curse,
A deadly sentence.
It is also a blessing.
We are nothing but time,
Lose it and you lose it all.

Again, it's a cliché, yet time is all you've got. Time is what your whole life is made up of.

Each day in this world is a blessing while you slowly descend into your end. Wishing your time away is to speed up the process. Time lost is never found. Once gone, it's gone.

Think of everything you do in life, including every action, emotion, and touch. Someday, you will do it for the last time. That is fine for certain things that won't be missed. However, knowing you'll never have a chance at other things is devastating.

When did you last talk to your parents? Kissed your spouse or hugged your child at night?

What if you knew that was your last time? Would you be crushed?

Completely. Certainly.

Contemplating such questions is an ancient Stoic gratitude practice called *"last time meditation."*

It puts things in perspective. It allows you to value your time, be grateful.

Without gratitude, you notice only what is missing. With gratitude, you celebrate what's there, *"reflect upon the blessings you possess"* (Maxim 7.27).

"Last time meditation" will help you to enjoy the moment you are in, and *"to learn to live in what alone is life—the present!"* (Maxim 12.3).

Look around. At those who are beside you. Think of what you're doing.

Realize that this may be your last chance at it all.

You'll feel grateful for being able to enjoy this moment. Live this life.

This is why you have to make the most of your time. Dream big, chase rainbows, build castles in the sky!

Time is too slow for those who wait and too fleeting for those who fear. Don't wait for better times; be resolute to live life *now*.

Or else, one day you'll wake up overwhelmed with a rush of sadness, realizing that your entire life has passed you by like in a blurry dream...

"I wasted time, and now doth time waste me."
William Shakespeare

It's your time now to create a life you value and love. Because someday, it will no longer be your time no more.

Life is short; death is long.
Don't wait.
Life is happening now.
Enjoy!
There's plenty of time to be dead.

Maxim

2.5

"At any moment, firmly, as a Roman and a man, resolve to do the work in hand, with scrupulous and unaffected dignity,

with justice, tenderly and willingly, securing respite from all distractions. This you will do if you perform each task as if the last thing in your life, rejecting every frivolous diversion, denial of the rule of reason, hypocrisy, self-love, displeasure with allotted share. You see how few things a man need master in order to secure smooth flow of a godly life—for of a man that holds to these principles, the Gods require no more."

Marcus Aurelius. Meditations 2.5

"As every divided kingdom falls, so every mind divided between many studies confounds and saps itself."
Leonardo da Vinci

Multitasking indicates a restless mind doing many things poorly at once. As an illustration, Albert Einstein suggested this:

"Any man who kisses a pretty girl while driving safely just does not give the kiss attention it deserves."

A good job and clear thinking demand a fluid focus and careful attention. Today, we call this "*mindfulness.*" The Stoics used the term "*prosoche*" (attention): to be at your best in any circumstance and life scenario, you must be present and aware of every step *(Maxim 12.3, "Learn to live in what alone is life—the present").*

Imagine you are among the flowers of a sunlit meadow, walking shoeless in the grass, when suddenly you enter an area with bits of broken glass. You're walking extra carefully now, watching your every move like a hawk to make sure you don't hurt yourself—that's the attention the Stoic aspires to pay to his every action.

It's not a tense state. You become fully present, developing a gentle focus that enables you to see the tiniest details without losing the simple joy your bare feet evoke. It's an elegant state where progress is fluid and smooth, the whole life just appears to flow.

This focused, elegant attention and continual self-observation are crucial if you actively want to align all your actions with virtue. How can you ensure that you are acting virtuously if you are not even aware of every act? As you let your thoughts drift away, your acts become mindless, you fall into folly, and you give up your chance for happiness (*eudaimonia*) because you are not at your best at that moment.

> "*Attention (prosoche) is the fundamental Stoic spiritual attitude. It is a continuous vigilance and presence of mind, self-consciousness which never sleeps, and a constant tension of the spirit.*"
> Pierre Hadot, *Philosophy as a Way of Life.*

Even with this "tension of the spirit," it's probably not always possible to be completely faultless, but you must try at least and—as Epictetus put it—be content if, by never remitting this attention, you shall escape at least a few errors.

Elegance is reached when all that is superfluous has been discarded and a human being finds simplicity and concentration:

> *I want to do a certain thing*
> *in this world,*
> *and I am going to do it*
> *with unwavering focus!*

Portrait of Julius Caesar, Aegidius Sadeler (II), after Titian, 1624 - 1650

Young Man with a Skull, Lucas van Leyden, 1517 - 1521
The young man wears an ostrich-feather hat and holds a skull, half-hidden in his cloak.

Maxim 2.11

"ACT, SPEAK, AND THINK LIKE A MAN WHO COULD TAKE LEAVE OF LIFE AT ONCE."

Marcus Aurelius. Meditations 2.11

Death is a stealthy predator, lurking in the shadows of life. Ready to strike at any moment, whether it be this afternoon or the next morning; as you go to work or as you go to bed... One day, the ordinary course of life will be terminally punctuated by the extraordinary full stop of death.

Is it possible at all to live each day to the fullest? To relate to others as if there were no tomorrow?

Yes, it is.

If only you would act as if each of your life's actions were the last.

You can't let even one of the priceless moments of life slip by lightly if you are looking at things from the perspective of death, *"as one who's dead, whose life till now lived and gone"* (Maxim 7.56).

If you believe, like Marcus, that moral action and pure intent are the only good, then the death outlook will transform the course of your actions and thoughts in this very instant.

The thought of death infuses every life's moment with beauty, gravity, and meaning. Enabling you to live the present moment with such intensity and passion that, in a sense, your whole lifetime be contained and completed within it.

This is what it means to "act as if each of your life's actions be the last."

When you feel like slipping into something empty and frivolous, like binging on TV or aimlessly scrolling through social media, remind yourself that what you now see, or think, or do could be the last.

What would you really want it to be?

When you feel like flames of anger threaten to ignite, towards a friend, or a stranger, remind yourself, next move could be the last.

What would you really prefer it to be?

When you say you are "too busy" to your loved one or your child, remind yourself, next thing you think, or say, or do could be the last.

So what would you like it to be?

Would Marcus' maxim always work? Most likely not—you are a mere human! But bearing in mind the fact that your next thought, an action, or a phrase could be your last one and placing it in the context of imminent death, will definitely help you make more reasoned choices.

Do not shun death. Keep it close to you. Like Mexicans do. Who are tight and intimate with the dead not only on the Day of the Dead,

"Todos somos calaveras." ("We are all skeletons")

A quote from Epictetus, which Marcus Aurelius references in Meditations 4.41, works the same:

"You are a poor soul, saddled with a corpse."

Say often to yourself:

I face the impermanence of life, the fleeting nature of time.
In the shadow of death, I find myself more mindful of the choices I make.
I know that at some point one of my words, thoughts, or deeds will be my last one.
I will make it a good one.

Maxim

3.4

"Do not waste what is left of life worrying about others,
save only when your goal is the common good."

Marcus Aurelius. Meditations 3.4

Yes, too much worry about what others may think would prevent you from conducting your business of passion.

Albert Einstein allegedly had only two rules of conduct. The first one: *"Have no rules"*. The second one: *"Be independent of the opinion of others."*

Why? The thoughts of others could become your cages!

"Care about what other people think, and you will always be their prisoner."
Lao Tzu

It's not the risks of trying something new that usually hold people back, but the harsh verdicts from an invisible jury they've formed in their heads. Still, it's not your job to be what other people want you to. It's their error, not your failure. Your main goal is not to fail as a human, to *"do nothing but leading to some social end"* (Maxim 12.20).

The vital good in life is anchored within the area of your control, in your own actions and judgments, and everything else, other people's opinions included, is effectively 'indifferent' with regard to the ability of living a good life.

"For why neglect your proper duty in wandering about what such a one is doing, why, what he is saying, thinking, and devising; or any other vain imagination that keeps you from allegiance to your own reasonable self."
Meditations, 3.4

Just work to meet your own standards in full accordance with your values *"and fear not that life will ever stop, but that you never start off living in accord with nature"* (Maxim 12.1).

Doing your best job then becomes your steady source of self-respect, reducing your reliance on the unsteadiness of other people's thoughts. Moreover, when you realize how seldom, if ever, other people think of you, you'll start to care dramatically less about what they do.

Now, aware of the fact that the jury is invisible, assured that you're following your human nature, and that most people are mostly focused on their own lives, go do whatever it is that makes your soul shine.

I am a flame,
unapologetically burning bright.
I will not dim my light for the sake of another's sight.
I act for myself,
my will, my way, my right.

The Virtuous Man, Crispijn van de Passe (I), 1589 - 1611
Man is diligently at work, surrounded by the tools of the artes liberales and the artes mechanica. In the background, we see the choice of passions, represented by a couple making love.

Pride, anonymous, after Frans Floris (I), 1575
Personification of Pride in the form of a seated woman, who looks into a convex mirror in her hand. Behind her is a drapery. Next to her is the peacock as her attribute.

Maxim 3.5

"No overdress of thoughts in finest language.
Too many words, too many deeds to be avoided."

Marcus Aurelius. Meditations 3.5

Most of the ancient Stoic texts are very accessible. And Marcus Aurelius was particularly known for his skill at conveying the beauty of things in a few choice words.

He was concise on purpose. Explanations with too many words and fancy language often hide a lack of knowledge. The audience is expected to *"be awed by the grandiose phraseology and the air of sagacity, their doubts silenced by the fear of appearing unenlightened in the face of such verbal extravagance."*

The highest knowledge is short and clear. It is the road to truth; the rest is just show, empty talk, and grandiose language, giving no knowledge:

"A sophistical rhetorician, inebriated with the exuberance of his own verbosity and gifted with an egotistical imagination that can at all times command an interminable and inconsistent series of arguments, is merely drawing out the thread of his verbosity finer than the staple of his argument."

In brevity, the truest art
Less is the language of the heart.

Wrong belief alienates the world from the truth, Dirck Volckertsz, 1575 - 1581
A man with a fool's hood is the personification of the Foolish World (Mundus Fascinatus). He is enchanted by False Belief (Opinio), a woman with two glasses on her forehead. She reads magic formulas from a book. A magic circle has been drawn around Foolish World with the symbols of the seven planets, as well as a holy water brush, a convex mirror and a burning candle. The personification Truth (Veritas) tries to help Foolish World, but is attacked with a sword by the enchanted Foolish World. In the background, Justice plunges down a rock.

Maxim

4.12

"*Always.* BE SET TO CHANGE YOUR COURSE, IF ONE BE NEAR TO SET YOU RIGHT,

and guide you out of some vain opinion."

Marcus Aurelius. Meditations 4.12

People want to be right, right? Paradoxical as it may seem, but for the ancient Greeks it was the other way around: to be wise, you must be able to find pleasure in being proven wrong.

In a philosophical dispute, he who is defeated gains the most, since he learns the most. — Epicurus, Vatican Sayings, 7

He who wins the dispute gets nothing except some kudos, maybe. But he who loses gets knowledge and comes closer to wisdom and truth.

The Greeks believed benevolence and truth were divine qualities. A mortal can emulate this second quality if he is willing to set aside his ego and accept the truth that someone else has declared. Zeno, the father of stoicism, claimed that giving up one's purpose if persuaded by reasoned advice is the greatest victory over oneself. This act requires courage and makes you godly strong.

And godly free.

Some people may think admitting you're wrong while another is right is a sign of dependence and weakness.

The truth is, changing your mind to please others is a weakness and slavery. Whenever you renounce reason and the truth, you get enslaved to things external and the will of others. Ultimate freedom would be doing what you think is right and true, no matter how much it hurts your pride or ego.

"Remember: to change, and to follow him who sets right are free acts alike; for it is your own action, done in accord with your will, and your judgment, and mind."
Meditations, 8.16

Other people may say that the powerful are always right.

The opposite is true.

It is through admitting your errors while genuinely recognizing you were in the wrong that you get strong and free. A strong man has just as much potential for error as a weak man. The key distinction is that a strong man would admit his mistakes, laugh at them, and learn. This is how he gains strength.

Marcus Aurelius was the world's most powerful man at the time, and he sought to be delighted rather than offended if someone proved him wrong. It shows in Marcus' efforts to model himself after his adoptive father, the preceding emperor, Antoninus Pius.

In Meditations 6.30, Marcus tells himself to be *"in all things the disciple of Antoninus"* and carefully lists the qualities he most admires in him (which seems to be a recurring spiritual practice for Marcus; *Maxim 4.38, "See clearly what governs mind of wise"*).

Among many other virtues, Marcus points to *"his tolerance of frank opposition to his opinions and delight when anyone indicated a better course"* (Meditations, 6.30).

On top of that, admitting mistakes builds more trust than appearing flawless. Seek to seem flawless, and others will seek to expose your flaws. Admit your mistakes, and others will seek to assist you be flawless.

Take this lesson to heart and apply Marcus' mindset to every action you take.

"Just do the right. Regard of nothing."
Maxim 6.2.

Maxim

4.24

"Do few things — if you seek cheerful calm.

Or, better put, do what is a natural necessity; what reason of a social creature proves, and as it proves. For this brings calm. Doing few, doing well."

Marcus Aurelius. Meditations 4.24

Marcus is referring to Democritus' saying here. Epicurus captured its spirit as follows: *He who intends tranquility must avoid doing many things, in public and in private, and in what he does must not undertake what exceeds his strength and nature.*

We are truest to our nature in calm. Like water. To become calmer, do less. But Stoics, at the same time, valued hard work. Being hard-working was considered a virtue. As well as moderation. So, how should we do less and still be productive? The simple answer is: *Do not do unnecessary things.*

"Most things we say or do are not necessities; get rid of them, and you will gain both time and calm."

Meditations, 4.24

With so many options for entertainment and gain all around, we cannot help but always fear missing out on the chance for fun or money. Daily opportunities are just too many to seize them all without exhaustion. As a result, we typically end up drawn, like moths to a flame, into the clutches of busyness, caught up in all kinds of things that lead nowhere, just wearing us out, draining us of energy and time. Not that we shouldn't be active. It means we should carefully choose where to place our focus and how to expend our most precious hours.

Recall your days. Think of the siren song of social media, its endless scroll, its endless drama. Think of Netflix binges and video game thrills. All these lure you in, fight for your energy and time, *"which, if you don't dispel the mists within, will pass away, and you along with it; and chance will never more return"* (Maxim 2.4, "Recognize that limit of your time is set").

Still, somewhere deep inside you, there must be a voice telling you quietly what you truly need and what you needn't be doing. It is crucial to learn the skill of saying "no" to activities that do not feed your soul.

"Thus, on each occasion, ask yourself, Is this a necessary one?"
Meditations, 4.24

It's crucial to figure out what's essential. Ask yourself deeply and honestly. *Do you really need it? Will it make you more tranquil? Or happier? Or does it cause you stress?*

Keep asking yourself, every time, every day.

Marcus Aurelius had one of the busiest jobs imaginable. Still, he made time to slow down, to sit and write in his journal, meditating on wisdom, death, and tranquility. For him, it was essential.

"Wisely and slow, they stumble that run fast." Shakespeare. Romeo and Juliet, 2.2.

When you cut out the noise and focus on what matters, your mind finds calm, and true progress takes form in the essential actions.

The sad truth is that most of the things we want, automatically react to, or buy into are not essential. They do not make us happier. Rather, they are the reason we aren't happy, yet. That is why, *"at any action ask yourself, what does it hold for me? Shall I repent of it? A little while, and I am dead; all these things are gone"* (Maxim 8.2).

Doing few means not only avoiding unnecessary things but also doing things well, with all your heart. Life is about quality, not quantity, Seneca said. So long as you don't waste it, there is enough time in life. It's like a play: what matters is the excellence of acting, not the length.

Do less. Calmly. Better.

Moderation, Jakob Frey (I), after Domenichino, 1721 - 1725
Temperance (Ripa), one of the Four Cardinal Virtues.

The Four Elements, Jacob Matham (attributed to), after Hendrick Goltzius, 1588
The four elements Earth, Water, Air and Fire, personified as two female and two male figures. In the left foreground, seen from the back, is Earth (Terra) with sheaves of wheat in his hand. Also, in the foreground is Water (Aqua) as a female figure with a ship on her head and a stream next to her. In the sky between the clouds sits Lucht (Aer) with a chameleon on her hand. Rear right a male figure representing Fire (Ignis), sitting on a dragon and a crown of flames on his head. Behind his head the sun.

Maxim
4.36

"Ever behold as all things come to be through change;
get used to the idea that nothing so loves the Nature as changing things that are, and making new in their likeness."

Marcus Aurelius. Meditations 4.36

When Marcus quotes Heraclitus' words that no one steps in the same river twice, it's possible he's not just being metaphoric. He could have meant his days spent by the river, wide and deep, in Carnatum or Aquincum, reading, thinking, and growing wise. As Marcus sat by the rushing stream, contemplating the fickle nature of reality, he could've gazed upon the acorn at his feet, considering the journey it would undertake. From its humble beginnings as a tiny seed, it would grow and transform into a grand oak tree with branches stretching towards the sky.

Everything in life is flux, constant change, a never-ending flow. We too are changing, day by day. As time and life, days come and go.

Change comes to us in endless ways. Our bodies age, become ill, or heal. Our hair grows and then it fades. Lines crease our brows, and shadows linger beneath our eyes. We make new friends, move on, and say goodbye. We fall in love, and then we fall apart. We learn the truth of change, only to forget it once again.

This is the way of life. We must embrace it. Change is the constant companion that touches us all.

Look upon the world around. Life is built on change, from darkness to light, from life to death. From childhood's joy to autumn's chill. Change is the way of the cosmos. A dance of transformation, never causing harm to the grand design.

"Termination is but transformation, and that is a Nature's delight."
Meditations, 9.35

Transformation is woven into the very essence of life, and the loss or termination of anything is neither good nor bad. It is our perception that renders it so. Our suffering is simply the fear of change and the trials it brings. Inability to fully embrace the shifting tides, the beauty, and the pain of existence.

"If you are pained by anything external, it's not the thing that troubles you, but your verdict on it. It's upon you to annul that right now."
Meditations, 8.47

As you navigate the twists and turns of life, remind yourself that change and transformation are the very font of all that's good. Without transformation, nothing could exist, for it is only through change that anything and everything is formed.

"All things by nature change, decay, and pass away, so that other things be born in their place, in their turn."
Meditations, 12.21

The journey of life presumes transformation, a process of shedding old skins and embracing new beginnings. The cycle of birth and death drives the evolution of all living things. Change can be jarring, but without it, there is no growth.

It is better to embrace the beauty and struggles of life, *for life with twists and turns is sweeter still than one with no pulse.*

But why, when change is so natural, do we still struggle to embrace it? Why do we cling to what we know, even as it shifts beneath our feet like sand? Why, even when our reason tells us that change is a constant, do we so stubbornly resist it with all our hearts? Why do we exhaust ourselves trying to evade what is, ultimately, inevitable?

The answer is: *we enjoy control.*

Control is what makes us feel safe. Change is the thing that takes it all away. It's the unknown. Control is like a steady hand on the reins of our lives. Change, on the other hand, is the wild horse, galloping beyond our grasp.

Control gives us comfort—a warm fire to huddle around in the cold. For when we hold sway over our surroundings, we can eliminate all threats, all challenges, and all discomfort. However, most control is an illusion, a mirage we chase in vain. In reality, only two things are truly under our control: our thoughts and actions. All else fades, may be taken, lost, or is beyond control.

Whatever your will, life will bring changes beyond your command. Changes, unstoppable, will come your way. Either you rise to meet life's challenges, or crumble. That is the choice. Whining is useless *(Maxim 8.50, "Complain not")*.

You can only remind yourself, life is a journey of change and transformation, like a river, ever-flowing. Resist not. Accept and go with the flow. That is the best for you. Accept the things you cannot change, take charge of what you can. To accept is to embrace the path to a life of joy, fulfillment, and serenity.

So, you remember always: *change is natural, it is the very breath of Nature, its delight. The rule, not an exception.*

Do not be fooled by the illusion of stability, wherever it appears to exist. All things must eventually give way to the unyielding force of change. Even the stillest of stones will eventually crumble and give way to something new.

Embrace the changes and dance with them. Live life to the fullest, in all its ever-shifting glory.

To resist is futile.

Instead, you could treat change as a chance, a blessing in disguise, and try to "*smartly adapt what was granted*" *(Maxim 8.32, "Arrange life act by act")*.

Or you could try to find a new perspective.

When you feel that the moment is pulling you in, take a step back and see it from above. Consider what this moment really holds as if you were in the sky *(Maxim 7.48, "Take a bird's-eye view")*.

Or try to think through the eyes of the Sage.

Change, viewed through the eyes of the Sage, could bring to mind the proper course of action or emotion that would fit the moment best *(Maxim 4.38, "See clearly what governs mind of wise")*.

Whatever you decide, do not forget:

If you're happy with the way things are,
Then brace yourself. For this shall pass.
Unhappy?
Then have faith. This, too, shall pass.

Like the tides that come and go,
Change is the only constant flow.

Ancient busts of Socrates, Paulus Pontius, after Peter Paul Rubens, 1638

Maxim

4.38

"See clearly what governs their minds,

see what the wise indeed avoid, what aspire."

Marcus Aurelius. Meditations 4.38

To be a sage, the Stoics say, is our ultimate aim. But it's just an ideal, a lofty goal that few may attain.

The Stoic sage is kind of an idealized mental construct, a concept of perfection. Even Socrates, whom the Stoics idolized and often used as an example, was not considered a sage. For Stoics, we are all prokoptons, on a journey towards becoming sages. But before you can sculpt yourself into your ideal self, you must first recognize that self.

What does it mean to recognize your ideal self? The ancient philosophers developed a technique for this called "Contemplation of the Sage." You pick a person whose Stoic traits you admire as a role model. It could be a real figure like Marcus or an imaginary blend of traits and virtues.

Having a guide, a wise and noble sage, would help to shape your actions as you travel through this world. The thought of being watched by your ideal mentor should make you strive be true and honest, acting as if you were to answer to your perfect guide, being accountable for every move you make: *"I am watched by my ideal mentor, whose eyes seem to see into my soul. I act as if I must answer to them, for I know that they expect the best from me. Their expectations are a burden but also a source of inspiration. I will do everything within my power to live up to their expectations, for they are my guiding light, leading me towards my ideal self."*

When you have a sage to emulate, you could split your contemplation into writing and imagining exercises.

1. Writing.

The whole first book of Meditations is, in fact, a tome of reverence. Marcus honors those who have shaped his being. Book 1 begins as follows:

"From my grandfather Verus, good will and even temper."
Meditations, 1.1

Reflect upon the traits you admire most in those who surround you, much like Marcus did in the opening of his "Meditations." This by itself could bring encouragement and motivation. Contemplating a person who completely embodies a virtue is also a way to work towards embodying that virtue yourself. It proves the idea that the excellence you aspire to can, in fact, be attained.

Jot down a few key traits of each of those you admire to better envision the qualities defining them. This practice allows to keep their influence ever present. In times of trouble, turn to their wisdom, visualizing their virtues to guide you through.

Don't put it off; write your own Meditations Book I now!

2. Imagining.

Imagine in vivid detail the person you look up to. Your Stoic ideal is steady and true in all they do, unflinching in their daily routine:

The Stoic sage awakens, bodies rested, mind calm, and drawn to the day ahead with hope and grace.

A nasty word is thrown as they walk down the street, but they remain undaunted, their spirit ungrown, their peace unchallenged by the world's turmoil. Their confidence blooms, no need to rehearse fierce responses to the day's unknown woes. They trust in themselves, the moment will reveal how to react with poise.

Amidst the clamor of the office, a dispute arises, yet they remain unruffled, serene, addressing the concerns with ease.

They don't see darkness in the hearts of others and don't take things to heart. They forgive with ease and do not dwell on conflicts long since past. Envy is foreign to them.

They fear not the future, for they know that the forces that have brought them thus far will continue to guide them. Material possessions hold no sway over them; they know those are but passing things, and true wealth lies within.

They value the finer things but still find joy in less, but dear.

They tend to their physique, pumping iron, staying fit. They fuel their bodies right, but vanity is not permitted.

Worrying for others' well-being, they act to assist, but don't fixate. The burden of the world they do not shoulder.

See yourself, in your own mind's eye, acting like this ideal person repeatedly throughout the day. As often as you can, let your actions mirror those of this sage.

As friction greets you in the day, ask yourself, what would your sage suggest to do? How would they approach this obstacle? And how would they perceive your way of handling it?

Or more specifically:

"What would Epictetus do?"

"What would Seneca's actions be in these circumstances?"

"Marcus, job gone, love fled, sick and suffering—how would he carry on?"

On the other hand, the Stoics were lovers of wisdom, "philo-sophers," not gurus to be worshiped. Even the founders of Stoicism were not perfectly wise, by their own admission.

Stoicism was founded by Zeno, and stoics were also once known as Zenonians, just like the ancient schools of the Epicureans and Pythagoreans, named for their revered founders. But the Stoics abandoned that name, opting instead for *Stoa Poikile*, the marketplace where they would meet in Athens (the large columns that encircled these markets were called *stoas*).

With this in mind, you could aim to compare yourself to the person you dream to be, not some imaginary being.

If you had the courage of a falcon, what would you do? Would you dive and conquer? Or would you turn and fly away?

This would allow you to focus on the present you, while letting go of envy and jealousy towards others. It may well be the only person worthy of surpassing is you yourself from yesterday.

Stoicism is a philosophy, not a faith. Each must find their own truth.

So, aspire to see clearly what governs wisemen's minds.

Learn from them.

So, aspire to see clearly what governs wisemen's minds.

Learn from them.

Then forge your own path, don't fear error, embrace each lesson, and ultimately be yourself, not just a copy of someone more lauded, astute, or triumphant.

"If something's hard for you, do not assume that for a man it is impossible, but whatever is humanly possible, consider also attainable for you."
Meditations, 6.19

Death of Socrates, Jean François Pierre Peyron, 1790
Socrates, ready to drink the hemlock (...) (title on object)

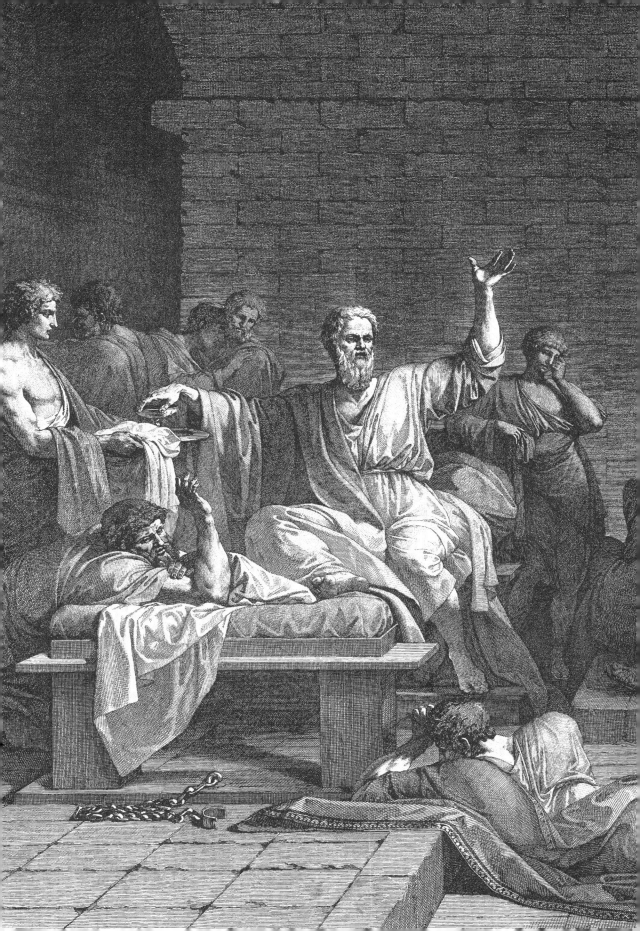

Maxim

5.23

"Think often of the swiftness with which the things that are, or come to be, sweep past and disappear.

Being is a river in a ceaseless flow; its action ever changing, its causes countless in variations. And scarcely anything stands fast, not even what's right here;"

Marcus Aurelius. Meditations 5.23

Only a fool would hold onto pride, pain, or anger for anything. All the displeasures of the moment are but a passing breeze, nothing born in this world can persist.

"Worlds on worlds are rolling ever
From creation to decay,
Like the bubbles on a river
Sparkling, bursting, borne away."
Percy Bysshe Shelley

The ultimate fate of all things, present and future, is to be engulfed by an ever-expanding, infinite past. Pride, trouble, anger, irritation—these emotions are the heights of folly for the one who comprehends that all of life is in a constant state of change and that all that appears will ultimately fade away.

Father Time Takes Life Away, Bernard Picart, after Carlo Maratta, 1693 - 1696
Father Time and the personifications of the four seasons row a sleeping man to an island. There Death opens a tomb.

"In the presence of eternity, the mountains are as transient as the clouds."
"Our lives are written in disappearing ink."
"We suffer because we cannot take the truth of transience."

Sure, this is all as clear as day... And this is the reason why these words must be echoed again and again, in a constant refrain. For this truth is like a feather on the wind, easily forgotten, easily blown astray. For it's so easy to fall prey to the delusion that joy or sorrow will last. To guard against the lie that all remains, keep this close to your heart:

Existence is a dance of in and out. Some things arrive, others depart.

Always remind yourself of this eternal game.

"All things by nature change, decay, and pass away, so that other things be born in their place, in their turn."
Meditations, 12.21

This cycle is nothing to be sad about.

It's just a fact to accept. a truth to hold on to.

Time is a cruel mistress, turning the endless future into a distant memory. We are but leaves on the river, carried away by the currents of fate. No anchor to hold us steady as the infinite past consumes us and the future slips away.

Look around. What should you prize in this world where everything seems to fade and fall?

A money's might? A power's allure? The glint of fame? The pleasure's sound? Or the others' hearts?

In the swift-moving stream, all these things are swept away along with us. If you chanced upon a sparrow, would you cling to it with love?

"Ambition is a meteor-gleam;
Fame a restless airy dream;
Pleasures, insects on the wing
Round Peace, th' tend rest flow'r of spring."
Robert Burns

Suffering comes not from fleetingness and change, but from the wish for things to stay unchanged when they can't.

To keep from becoming distressed by what happens and to let go of the burn of petty things that fade in the passing of time, recite this lesson like a poem:

"Infinity past and to come gapes before us, and swallows all things. How foolish then in all of this to puff, to strain, to fume, as though time and troubling lasted long!"

Meditations, 5.23

All things are brief and transient.

P.S.

"Outstare the stars.
Infinite foretime and Infinite aftertime: above your head,
They close like giant wings, and you are dead."
Vladimir Nabokov

Sentry Sitting on a Stone Block, Adolphe Mouilleron, after Joseph Nicolas Robert-Fleury, 1851 - 1862

Maxim 6.2

"Just do the right. Regard of nothing.

Cold or warm; heavy-eyed, or well-slept; cursed or blessed; dying or doing something else."

Marcus Aurelius. Meditations 6.2

Duty was the creed for the Stoics, hearts filled with a steadfast resolve.

Being a Stoic obligates you to yourself and your fellow humans to do what is right and be your best at all times. When you do what is right, give it your all, you prove your worth, and you should be satisfied with your efforts, no claims or regrets whatsoever *(Maxim 8.32, "Arrange life act by act")*.

But what is right and what is wrong? It can be hard to discern in life's complex and tangled affairs.

To act or to abstain?

To give or to get?

To aid and protect, or simply pass by?

When faced with a hard dilemma, your true self is revealed like a diamond in the rough. Brilliant and enduring.

No book of rules exists to tell you what to do exactly or when you must step in; it's up to you to weigh and measure, to use your best judgment in each situation.

In Stoic thought, the only things that bring true good are the four virtues that reside within human excellence: justice, courage, wisdom, and moderation's grind. Acting according to virtues is right; acting against them is wrong.

If you are truthful with yourself, you know deep down what is right and what is wrong. You simply feel it. It's easier than trying to define it. When you must ask the difference, perhaps you seek a way to validate your wrong.

In times of crisis, trust your instincts, use your judgment, follow your heart.

But do the right. In all types of situations.

There are three types of people, Marcus said. First, those who do good and expect something back. Second, those who give without asking, but view the other as indebted. The third type is just

"like the vine which has produced the bunch of grapes and looks for no reward beyond once it has borne its proper fruit"
Meditations, 5.6

This third type, like a hunting dog or a bee making honey, does not ask for anything except to move on to the next right act, and

"as the vine in due season produces new clusters again."
Meditations, 5.6

You have it in your veins, do good to others, do it for its own sake.

This is the moment to do what is right.

To seek wisdom, to exhibit discipline, to embody courage, to pursue justice—and not just doing no harm, but actively helping others—these are the things that matter, and doing them is right.

Fiat justitia, ruat coelum.
("Do the right thing even if the heavens fall," Lat.)

Maxim

6.11

"When torn in pieces by the force of facts, fall back at once upon yourself, and
DO NOT BREAK THE RHYTHM MORE THAN YOU MUST;
you'll be more master of the harmony if keep on going back to it."

Marcus Aurelius. Meditations 6.11

Although the waves of life may toss and turn, accepting the reality of change brings a sense of balance and a tranquil mind (Maxim 4.36, *"Ever behold as all things come to be through change"*).

Still, we're all surprised now and then, not just by big things but by the little ones too. The everyday occurrences that catch us off guard leaving us feeling unprepared and adrift. A small event, a coffee spill, can spiral into a day of anger and frustration, one thing leading to the next, and the next, until we start to expect only bad things to happen. We tell ourselves it's just an off day, but really it's a choice to follow the path of unintentional disruption.

To fall, to stumble, is just human. a natural part of being alive. No need to fall into despair. The wise soul knows this truth; they aim to recover with speed like a punching ball that bounces back whenever it's hit. Like finding the beat again—in Meditations 6.11 Marcus shows how we can reclaim our harmony by returning to our true selves after being thrown off balance.

After all, even the great masters of martial arts, who are known for their strength and grace, stumble and lose their balance sometimes. There is a tale of Morihei Ueshiba, the founder of Aikido, who was admired for his expert skill and poise. A student in the dojo, unsteady on his feet, once asked him how he was able to stay so perfectly centered, never once off-balance. "I lose my center, it's true. But I find it fast, so fast it's not seen by you," Ueshiba replied. Instead of running from the hurt of falling, seek out balance, find your center, your equanimity, as quickly as you can.

Equanimity is a beautiful word. From the Latin *aequus* (even) and *animus* (mind), *equanimity* is a "balanced mind."

Balance is beautiful.

Equanimity is a flawless harmony, a beautiful balance of mind.

"You'll be more master of the harmony if keep on going back to it," says Marcus.

It's all about one thing, really. How quickly can you bounce back and find your center when knocked down? This is the art of equilibrium, the still point of a turning world, in the midst of confusion and flux.

Aim to unfold this in your life as follows:

- Notice when you're off your axis,

- Stop and take a few deep breaths. Embrace where you are now,

- Accept that you are "off" and let the judgments pass. Feel the pain without hiding. Only through acceptance can you move forward.

- Ask yourself, *"What can I do to find my center fast? To feel again alive and clear?"*

- Act with love and understanding. One step at a time *(Maxim 8.32, "Arrange life act by act")*.

Balance is key. A tightrope you must walk. A daily ritual of finesse, in small things and the big.

If you stumble, instead of beating yourself up, just seek to quickly find your way back to equanimity, to move ahead with calm and grace.

Apollo, Pan and a putto blowing a horn, Giorgio Ghisi, after Francesco Primaticcio, 1530 - 1582

Maxim

6.13

"Make this your practice through your whole life, where regards seem overly trustworthy,

STRIP THE FACTS BARE,

behold their worthlessness, and so remove their gloried regalia."

Marcus Aurelius. Meditations 6.13

Have you ever viewed a fork as a spoon with holes that get in the way of enjoying the soup?

Or a spoon as a failed fork, with a single, obtuse prong?

A fun example, but seriously, while seeing things from a fresh perspective, we challenge the reality we blindly accept without truly analyzing it. Thus, inviting a deeper examination of the truth.

The Stoic analysis strives to parse each impression, peeling back its layers to reveal any value judgments lurking within. Then ridding it of all value, false or true, leaving only the core. The aim is to seek out the raw, unvarnished reality of the world as it truly is, stripped down to its essence, and to find within it a purity untainted by the biases of good or bad.

Marcus Aurelius aimed to see through the façade, trying to reduce everything that inflamed his desire to its fundamental elements, its bare constituent parts, the "worthlessness" of which would snuff out his cravings.

Still life of wildlife, C. Le Coq, after Dirk Valkenburg, 1840

"As in regard of roast meat and other dishes the thought strikes you—corpse of a fish, or bird, or pig; or that this wine, Falernian, is but a grape juice; the robe, purpleedged, a lamb's wool dipped in shell-fish blood; or making love—the friction of the private parts, a sticky liquid, following the spasm."
Meditations, 6.13

Try your hand at it; take a sheet of paper and put it into words. The things that currently hold some sort of sway over you. Pen the common portrayal of the object as it claims to be. Then "remove their gloried regalia," as Marcus would say, and write what it really is. Read it again with detachment, and let the hold it had on you disappear into the ether.

Chances are, you'll see the car you were longing for—the perfect blend of sport and style, advertised to make you shine—is just a blend of steel and polymers, electronics and fluids, no different from any other car. Then why drain your limited funds into things deemed valuable by a cruel trick of the mind, all for the sake of a deceitful illusion of status?

"Strip off the coverings, look at the causes."
Meditations, 12.8

For
"nothing produces the greatness of the mind like the ability to look into each thing in life in a methodical and truthful way."
Meditations, 3.11

Maxim
6.31

"Sober yourself, recall your senses.

Shake off the sleep again and realize they were mere dreams that troubled you."

Marcus Aurelius. Meditations 6.31

Marcus may have meant it straight. The Romans drank a lot of wine, and it was a staple in every soldier's provisions. And for Marcus, the opium may have added to the hazy feeling of the morning after. For long periods, Marcus took his daily dose of "theriac," a supposed cure-all, to ward off poisoning from foes. It made him drowsy with its hint of opium or mandragora, so he cut it out. But without sleep, he brought it back.

In the pages of Meditations, it is clear that Marcus kept his notebooks close to his bed, like a trusted friend to whisper secrets to. This particular entry was likely written in the stillness of night, after a nightmare had woken him from slumber.

But there's more to it all.

The things that shake us are mostly born from our imagination, or how we read the occurrences in our lives. The more we linger in these imagined worlds, the stronger our fear becomes. We spin our wheels, caught in the thrum of fear, imagining horrors that might come.

But it's not the reality itself that's frightening at all; it's the visions in our minds, the monsters we call, that are making the reality seem terrifying.

The Dream of Reason Produces Monsters, Francisco Goya, 1799
In the image, an artist, asleep at his drawing table, is besieged by creatures associated in Spanish folk tradition with mystery and evil. The title of the print, emblazoned on the front of the desk, is often read as a proclamation of Goya's adherence to the values of the Enlightenment—without Reason, evil and corruption prevail.

Fear is often a chimera, a fanciful illusion. Our mind is a canvas, free to paint. A playground of endless creation where anything can be brought to life. It's a gift. But when we let go of reason's reins and start to believe that our imaginings are true, we leave imagination's ground and tread the path of self-destruction. Reality becomes a grinder where dreams and wishes crumble into dust with fear and pain.

Life becomes just that—a dream, a dread.

But only the fear itself should be feared. We must awaken and end this madness' race.

How?

The typical response to fear is to flee trying to fill our minds with thoughts on other matters. But this is perhaps the least effective technique. The fear that lurks within the mind grows stronger if ignored.

Here, Marcus offers another way to handle it.

The root of most fears is the unknown. But as rational beings, we can confront and conquer it.

"El sueño de la razón produce monstruos." ("The Dream of Reason Produces Monsters," a Spanish proverb)

Embrace reason and get to know what you fear. With understanding comes control of excessive emotion.

To tame the fear, inspect it with reason and poise until it becomes a familiar presence. Ask yourself, "Is the thing you fear real, or just a figment of your imagination? Is it a dream or the dreaded creation of your own mind?"

If you learn to let go of what you can't control and recognize:

that your experience is shaped by the beliefs you've agreed to, just as a dream is shaped by your mind *(Maxim 6.13, "Strip the facts bare")*;

that life, like a dream, is just a collection of stories. Every story you experience can be molded and shaped to fit a truer perspective *(Maxim 7.48, "Take a bird's-eye view")*;

that change is the only constant in the Universe. Each moment, you're a fresh being, seeing with new eyes *(Maxim 4.36, "Ever behold as all things come to be through change")*;

that judging the moment causes pain, dwelling on the past or fearing the future brings no gain. But these views of reality are mere choices you choose to make *(Maxim 12.3, "Learn to live in what alone is life—the present")*,

you'll see at once, your fears are but an imagined creation, a dream, a mirage of the mind.

And then,

cleansed of hope and fear,
Freed from what once held you captive,
You find courage,
In the eyes of powerlessness, smallness, and doubt.

This is Marcus' call to life from sleep and death.

"Now awake again, look on the things just like on those dreams."
Meditations, 6.31

Maxim
6.38

"Often
MEDITATE UPON
INTERCONNECTION OF ALL THINGS
in the Universe and how they relate to one another."

Marcus Aurelius. Meditations 6.38

Here Marcus speaks of *"sympatheia,"* an affinity of parts intertwined in an organic whole.

We've been told to believe we are unique. Superior. The chosen few. Instead of unity, we focus on rivalry, on being the greatest and standing alone, as if we don't stand on the shoulders of those who came before us.

But In this vast Universe, all things are bound. You are linked, in one way or another, to everything and everyone. To each living thing, in a biological sense. To the earth, in a chemical way. To the entire Universe, at an atomic level.

You are but stardust.

In the moment of the Big Bang, all the atoms in the Universe were smashed and fused together into a single tiny dot that burst and spread, scattering through the vast expanse of space. The atoms that now make up you and those that once formed Marcus Aurelius were surely joined together then, colliding and merging through time. And when the solar system crumbles, your atoms will fly off into the endless void, to live eternally as mass or energy, being part of this inexplicably elegant Universe.

Just sense it.

Perhaps you have experienced that feeling of awe and wonder as you stand on the shore, gazing out at the vastness of the ocean, or on the rim of the Grand

Unity (Concordia), Cornelis Cort, after Frans Floris (I), 1560
The personification of Unity sits under a tree and with her finger lures a bird of prey that sits on a branch of the tree. With her other hand she pulls down the branch on which the bird is sitting.

Canyon, marveling at the grandeur of the endless depth, or looking up at the glittering myriads of stars in the night sky, wide and wild. These grand, natural wonders have a way of making us feel small and insignificant, yet also a part of something bigger.

It's the sensation of eternity, a feeling of something unbounded, an indelible bond with the world as a whole.

This oceanic feeling comes when you let yourself have a zoomed-out perspective, *"to take a bird's-eye view of earthly things, to look down on them as if from somewhere above"* (Maxim 7.48).

This feeling, called by Buddhists "oneness with the Universe," happens when you let go of your ego, your bubble, and see the world as one interconnected whole, understanding the interdependence of all.

To the Stoics, the whole becomes a new form of life. We are all a part of this one single organism that is the Universe.

This concept, *sympatheia*, is the basis of Stoic philosophy.

"For In a way, all things are interwoven, and thus have sympathy for one another."
Meditations, 6.38

All of us are here, connected to one another. It's our nature to be social creatures, intertwined. Man is by nature a social animal. As Aristotle said:

"Anyone who either cannot lead the common life or is so self-sufficient as not to need to, and therefore does not partake of society, is either a beast or a god."

We are all the same, bound by a common thread. We inhale the same air, harbor the same desires. Our dreams intertwine, our humanity a unifying force.

No matter who we are or where we come from, we all seek the same. We've all known the pleasure of play and the terror of judgment. Emotions, in all their hues, are something we all share.

In realizing our common bonds, we are brought closer to each other. "*Sympatheia*" whispers in our ear that we are not alone, that we are part of a larger whole, a greater good that demands our loyalty, above and beyond our own egoistic pursuits.

It's a reminder to reflect on tender things. Who am I, what do I stand for? What is my role in this world? Are my problems worth fighting for? Am I but a speck of dust? In the grand scheme of things, are my worries and my lust just fleeting flutterings?

Once this understanding blooms within, your gaze shifts outward, and the bonds begin to tighten as you see how all is entwined. With this new perspective comes a greater weight, a sense of purpose, to be and create for something bigger than just yourself. A new role to play in the world, to make it whole.

In Marcus' mind, understanding how we are all linked, all one, dependent on each other, prompts us to be good and do good for one another.

"Life is short, one fruit of earthly living is righteous attitude and action for the common good."
Meditations, 6.30

Being a cosmopolitan, a citizen of the world, the Universe, is the natural outcome of *"sympatheia"*. It matters not the hue of skin or faith you hold, your gender preference, or any other differences. What matters is showing love and respect to others, for this is the key to fostering the well-being of all. Ultimately, our differences are trifling. What counts is that we're all the same. Our universe, our planet, and our humanity are all connected by the beauty of *"sympatheia"*.

"For we have come into being for co-operation, as have the feet, the hands, the eyelids, the rows of teeth, upper and lower."
Meditations, 2.1

We are all connected, our lives entwined. So let kindness be the guide for your actions and your heart. By prioritizing compassion and care, we can create a world that is peaceful and fair.

So be good to your fellow man, for this is the right thing to do.
And you can.

Maxim

6.50

"Remember, too, that you

SET OUT WITH A RESERVATION,

*and aiming not at the impossible.
What was your aim then? - Effort, such as this.
And you succeeded. What you set out to do is achieved."*

Marcus Aurelius. Meditations 6.50

What is it like to *"set out with a reservation (or reserve clause)?"* In fine, it's about being fluid, not being too tied to a certain result. Expectations are kept in check, except for those within your control. Yours is only the power to shape your thoughts and intentions. Thus, you steer yourself towards your chosen fate.

The *"reserve clause"* speaks of two things.

One, to try and give your best from the start. And two, to accept that fate may not align.

Success is not promised, but you must try and shine. Adding a caveat, a gentle reminder of fate, *"If nothing prevents me"* or *"Fate permitting."*

Fate happens, after all;)

If you can't change things, change your attitude.

"Whoever yields properly to fate is deemed wise among men, and knows the laws of heaven."

Euripides

When you detach yourself from the desired outcome, you find serenity, even when things don't go as desired. But those who believe they can bend fate stay slaves to their own minds, not even fate, and suffer endlessly when trying to resist the harsh reality.

When the outcome is not a crushing load, you can move ahead, unafraid to try. Just because fate has dealt you a less than perfect hand, you must not allow yourself to despair. You must play the cards you have been given and strive to make the most of them. Just focus on what you can control, the rest shall unfold as it should.

It could be argued that if you don't expect a result, you won't use all your resources to attain your goal. But remember, the aim of Stoicism is to live a virtuous life. A Stoic would put every ounce of their being into achieving a goal—not for the end result, but because it aligns with virtue and is the right thing to do.

Stoics value success over failure. But failure is not the end, rather it's an opportunity to continue on the path of virtue. If your goals elude you, like a Stoic, you must say, *"I may have failed, but my virtue remains."* You simply keep on living, virtuous and true. This is your goal. You did not aim at the impossible.

"And you succeeded. What you set out to do is achieved."
Meditations, 6.50

Fable of Cupid and Death, Dirk Stoop, 1665
Death and Cupid have traded their weapons. Death, in the guise of a skeleton, shoots old people, who dance merrily in a circle. Cupid shoots young people, who drop dead from the arrows. In the foreground is a grave. Illustration from Aesop's fable.

Each thinks his owl to be a falcon, Hendrick Goltzius, after Karel van Mander (I), 1590 - 1594
A man sits on the floor and looks through his glasses at the owl on his hand (thinking he sees a falcon). Behind him two jesters, each with a marot (jester's staff) in hand. The proverb depicts the idea that everyone thinks they have or are the best.

Maxim

6.53

"Practice really connecting to another's words, and do your best to
GET INTO THE SPEAKER'S MIND."

Marcus Aurelius. Meditations 6.53

To be heard—truly heard—is a desire shared by most of us. Finding an ear that truly hears is a rare and precious thing. Most people do not listen well.

Next time you converse, observe the way they speak. You'll find that all they do is talk about themselves. No matter what the topic is, they'll find something from their own lives to insert. They do not truly listen to the words, but prepare their voice for when it's their time. They ask a question, but don't wait for the answer. Eager to interpret, counsel, or relate experience of their own. Their tales of triumph may bring them joy, but others may not find them quite as sweet. It's egotistical and annoying, leaving others feeling far from joyous.

Accepting a voice within, that calls to speak straight away,
Hold your tongue, resist the impulse.
Let your presence to another's need to speak be your wordless reply.
Let silence be your speech.

The depth of your focus is the purest gift that you can offer to another soul. Be the one who listens most of the time, only adding words that enrich the exchange.

What is the speaker trying to convey?

This is the question to constantly ask.

By delving into the speaker's mind, you cultivate a talent vital to all human bonds: the skill to grasp and truly empathize with the emotions of another being, to see through others' eyes.

"In conversation, attend to what is said, and in action, to what is done. In the latter, look at the aim of every move; in the former, watch the meaning of every word."

Meditations, 7.4

Connecting with others, not acting for them. Paying attention. Letting them speak the most. Finding pleasure in listening. This is your never-ending, noble task.

There's a reason the words *"listen"* and *"silent"* share the same letters: in order to truly listen, you must first embrace stillness. And learn the art of being present in the quiet spaces between words, for it is in these moments of silence that true understanding is born.

"Knowledge speaks, but wisdom listens."

This is the art of listening,

Words, sparing.

Silence, rich.

Utter a few words. Absorb much.

Maxim

7.1

"What is evil? What you have seen time and again. So whatever befalls,
Have this ready to hand:
"I've seen the like before."

Marcus Aurelius. Meditations 7.1

Nothing new is left to find for the wise. Nothing is beneath the sun that hasn't been seen or said before. Nothing to get freaked out about. Only the unwise fret and fuss over things that constantly repeat. Forgetting what the wise remember always,

"That all that happens has always happened, and always will, and happens now everywhere. Just the same."
Meditations, 12.26

Through plagues and wars and upheaval's tumultuous reign, Marcus Aurelius looked for a calm within himself, reminding him that history was always filled with such, and thus he did not let it weigh more heavily upon him than it earned.

Allegory of Life, Giorgio Ghisi, after Raphael, 1561
Fantasy landscape at night with rocks, wild animals, different kinds of trees and a raging lake with sea monsters and a rowing boat. At the edge of the lake stands an old man with a beard, leaning against a rock and a dead tree. The old man looks across the water to a woman with a crown and spear in the foreground on the right.

SEDET AETERNVM
QVE SEDEBIT FOELIX

RAPHAELIS VRBINATIS INVENTVM.
PHILIPPVS DATVS ANIMI GRATIA
FIERI IVSSIT.

Don't let despair consume you. This isn't the end of all things. You are a witness to history, which has a way of repeating its ups and downs, for better or worse. It too shall pass (Maxim 4.36, "Ever behold as all things come to be through change").

"In brief, same things wherever, above and below, at every page of history, ancient, recent, and modern, same stories now in cities and homes. Nothing is new, all alike familiar and fleeting."
Meditations, 7.1

The past, a teacher, equips you for the future. Though you may face new trials, the world has seen them all before. This knowledge, as a shield, should guard you from surprise. Embracing both the past and future sets you free. Free to live the Stoic way. In the present (Maxim 12.3, "Learn to live in what alone is life—the present").

"In sum, life is short. Make the most of the present, with reason and justice. Temperance in hours of ease."
Meditations, 4.26

Maxim
7.27

"Dream not of what is absent as if it's now yours,
REFLECT UPON THE BLESSINGS YOU POSSESS,
and gratefully remember how you would long for these, if you had them not."

Marcus Aurelius. Meditations 7.27

Contentment lost,
Desire ever grows,
Happiness elusive,
As the wind that blows…

These lines resonate, don't they? The more things we gather, the less happiness we seem to find.

Why is it that joy eludes us as our possessions multiply and grow? How much do we acquire merely to keep up with the neighbors or because most men possess such things? Do we follow the crowd or truly desire these things? Are we truly our own selves or just shadows of society's expectations?

Why do we suffer and stumble, trapped in a cycle of social patterns and norms? Is it not because we follow conventions rather than using reason to guide us through life?

In a world that values ambition and possession, it can be hard to resist the temptation to join the fray. But perhaps it is worth it to carve our own way,

Personification of Joy of Life, Melchior Küsel (I), after Johann Wilhelm Baur, 1682

to be true to ourselves, even if it means standing alone in a desert of conformity and acquisition.

Ask yourself, *"What do you really want?"*

Question your needs and desires! You don't need to shop daily. No need to flaunt your life on social media to earn the praise of others in fleeting, ephemeral likes on screens and feeds.

You don't need much, just what you need. No more, no less.

And be at ease.

And let go of the rest.

He who doesn't find content with what he has would find no joy in what he longs to grasp.

"They are not poor that have little, but they that desire much. The richest man, whatever his lot, is the one who's content with his lot."
Seneca.

Why do we find it hard to embrace the present, the good, to see the beauty in our lives, to appreciate the people we should?

Count the gifts you've been given—the love, the joy, the peace within. Don't let your worries steal your light.

Focus on your blessings. Count them!

Recognize that within your control lie the answers to the questions posed at the beginning of this chapter.

For now, seize the chance to value the things that truly count. If your priorities lie with material things, which only leave one feeling void, now is the time to reassess and realign. To shift the focus.

It is not riches that bring true joy and contentment, but rather the love and appreciation of those around us, that fill the heart and nourish the soul. So cast aside the fleeting pleasures of material things and embrace the timeless beauty of human connection. It is in these moments of pure, unbridled emotion that we truly come alive and flourish.

To be human is a gift. The greatest blessings lie within, said Seneca. Keep this in mind, if troubled, what are the finest things that you possess. Just press your wrist and feel the beat, an absolute reminder of the gracious gift of life. And thank the stars above for all its tiny joys and wonders. Like air, cool and crisp, that fills your lungs, wet grass beneath your feet, the trees that dance to the wind's sweet song. Loved ones by your side, laughter with those who are dear. Deep talks that never end, getting lost in a book's embrace that transports and inspires. For simply living, here, now. Breathing, thinking, being alive!

"We are rich in gratitude; in complaint, we are poor."

Mark well this truth and keep it close:

Blessings that you take for granted fade like mist under the morning sun. Hold tight to what you have before it's gone.

Maxim

7.29

"Wipe out mental images.
Stop acting like a puppet."

Marcus Aurelius. Meditations 7.29

A murmur from a colleague, a glance, a stranger's passing comment, a newsfeed scroll, a storm brewing on the horizon—we are adrift in a sea of impressions, tossed and thrown, not daring to question the forces that pull us around. And leave us lost, uncertain, tried.

We are strings tugged by invisible fingers. The puppets, limbs at the mercy of those who pull. Life is a choreographed routine, a dance in a labyrinth of unquestioned impressions. And then death catches up with a miserable puppet, eyes fallen inside...

And there is an alternative.

"Perceive at last that there is within you something superior and more divine than that which raises passions, and moves you like a mere puppet."
Meditations, 12.19

Marcus speaks of what makes us, humans, unique and separates us from all the earth. Our minds, our ability to reason, to seek the truth and question our own worth.

When you tend to your impressions and focus only on the strings that you hold, on what is under your control, when you sharpen your awareness, you master the dance of your soul.

You become the Maestro of your own mind. The conductor of the symphony of self. The one who commands the rhythm's flow, invites each voice to join in life's harmony. The orchestrator of the inputs, transforming chaos into beauty's shape.

The one who makes life sing,
Life beautiful, fulfilled.

Want to know the secret to it? Master the art of control.

Your judgments are like jazz notes scattered on a sheet, and like a skilled musician, you can master them and improvise a tune of beauty in any given situation life throws at you.

The Maestro of the mind is the one :

1. Who conducts with a gentle hand.

He chooses to reflect, not rush to respond; each note is a measure of grace.

In his slow dance, he notices impulses, impressions, and judgments and steps back, allowing them to pass like waves on the shore without being swept away *(Maxim 6.13, "Strip the facts bare"; Maxim 8.49, "Tell yourself no more than what's declared by the first impressions")*.

2. Who conducts his life in harmony with values, pure and fine.

The Maestro of the mind crafts a symphony of self-discovery where values— pure notes—lead each step towards a life of purpose and autonomy.

In such a life, one is free from the yoke of others' control and can move forward with full self-direction *(Maxim 7.59, "Dig within. Within you is the fount of Good"; Maxim 10.29, "Ask yourself, if loss of this makes death a terror?")*.

The Maestro reads music well. His fingers dance across the page. You must know your rules well too *(Maxim 3.13, "Always have your principles at hand")*. Then, in the face of chaos, just cling tight to the guiding principles of wisdom, justice, courage, and restraint.

In the end, the Maestro's wand wields a magic born of heightened awareness. Knowing what he's doing and how to do it in each situation. Bring that awareness, and you can have that magic too.

Don't rush, let the waves of emotion pass before you. Observe them with an objective eye and a gentle heart, with your values in mind. Then act with intention and purpose.

To sum it up:

Act slow, think clear,
do what's right*,
avoid strings tied tight.

That's all you need to not be danced around like a puppet.

* *(Maxim 6.2, "Just do the right")*

Don Quixote Destroys a Puppet Show, François de Poilly (II), after Charles-Antoine Coypel, c. 1723

Fable of the Bear and the Bees, Aegidius Sadeler, after Marcus Gheeraerts (I), 1608
A bear at three beehives. He was stung by a bee and, in his anger, knocked over one of the hives. He is therefore attacked by the bees that swarm from the fallen hive. The moral of the story teaches that it is better to accept things patiently than to seek revenge and suffer more.

Maxim
7.38

"Fret not at things external. For they care nothing."

Marcus Aurelius, Meditations 7.38

This quote comes from Euripides. It continues as follows: *"but if a man rightly handles the things he meets, he fares well."*

People often get angry over trifles that hold no weight, yet remain blind to bigger things. Like how they lose the beauty of their days, their lives slipping through their fingers like sand.

And you?

You spill a glass of wine, and with it, your temper. How often do such trifles incite anger in you?

In trifles, you find a test to keep your calm. To handle the small things right is to develop the virtue to fight the major battles yet to come your way.

This is plain: In a stress-inducing situation, ask yourself: *Is anger necessary? Is my serenity worth the cost of this?*

Perhaps the answer is a gentle "No." Nothing is worth the loss of calm.

Calm is a fragile thing. A bird's nest in the wind, easy to dismantle. And when dismantled, it leaves you with nothing but the weight of your own misery. You become less reasonable and less effective in stressful situations. You become a thing that can be easily broken.

Mindful of this, learn to see the chaos of the world as a currency for your tranquility, an exchange for the peace of your mind. As was beautifully put by Epictetus:

"Starting with things of little value—a bit of spilled oil, a little stolen wine—repeat to yourself: "For such a small price, I buy tranquility and peace of mind."
Enchiridion, 12.

Say to yourself before reacting to what angers you:
"I buy the calm instead."

It's not about the suppression of emotions but a dance with them, a delicate waltz of observation and release. Observe them, feel them, but let them flow. Don't suppress them. True freedom is in setting them free.

The challenge lies in the recognition of emotion as it arises, in the fleeting instant before it takes hold. In this awareness, you find the space to intervene, to disrupt the habitual patterns of reaction, and instead, with discipline, to actually buy the calm and not react at all.

You can't control anything outside of you, so do not let the chaos outside control you.

Red wine stains on white silk might seem a disaster. But it's a mere suit. Let not the stains of circumstance mar the beauty of your heart. Smile, do what's needed, and move on.

If, at a later time, you find yourself in strife, complaining of the trivial, then pause. And whisper, with gentle grace and measured breath, the words of Epictetus' wisdom,

For such a small price,
I buy the calmness of the ocean and the serenity of the sky.

Maxim

7.47

"To watch the stars in their courses, as one who is revolved with them."

Marcus Aurelius. Meditations 7.47

There's a kind of magic in the way the world turns, seasons shift, stars ignite. Nature's endless refrain, like a soothing balm, brings healing to the soul. A promise that winter's chill will yield to spring, night to morning's light. The sun will rise, with planets spinning 'round in cosmic ties, and will attend to the ripening of grapes on Earth as if it were the only thing that mattered in the vastness of the universe.

To breathe, to be alive, is a journey unlike any other. A riot of color and sound, a dizzying carnival of sensation. Yet so many people drift through their days on autopilot, their hearts no longer singing with the joy of living.

Many eyes go through the meadow, Ralph Waldo Emerson wrote, *but few see the flowers in it.*

Stoicism is not just about surviving the hard times, it's about embracing the beauty and wonder of life. Marcus tells us how to take more out of our time here. Subtle shifts in perception, that's all it takes.

Marcus suggests to dwell upon the beauty of life, to watch the stars *"as one who is revolved with them."* It means to sync your pulse with the pulsations of the Universe, align your essence with the symphony of Nature's heart.

It means letting the stars guide your heart and the wind guide your breath. It means to become one with the universe and let its energy flow through you in a seamless harmony of the self and the cosmos.

A rhythm, a beat, a song of the Universe
Echoes in my heart, in my soul
Nature and I, in perfect concord
Our spirits, one, in perfect whole!

Have you ever indulged in stargazing on those hot summer nights, when the air is heavy with the scent of blooming flowers? Have you ever lain on the grass, in the dog days of August, alone or with a beloved one, and fixed your gaze upon the stars until the first light of dawn? Have you ever truly savored the beauty of life in those moments of seclusion and quiet?

Life is a riot of such wonder, so much beauty it can break your heart. It's everywhere, in everything, a leaf of grass no less than the grandest journey of the stars.

In nature, nothing is without flaws, yet everything is perfectly imperfect. A constant reminder of the world's endless beauty. Trees, twisted and contorted, grotesque yet beautiful, a twisted smile of Nature.

Colors, sounds, scents, all around.
Nature's symphony: birds' tweets,
The city's hum, the coffee's aroma,
The sun's touch, a caress,
the moon's light,
the rain's splash,
and love.

This is beauty.
This is life.

The marvels of Nature are endless, like the ways the wind can kiss your face. See it all. Walk with open eyes and let Nature's splendor fill you with joy.

Embrace the rain. Let it kiss your face and drench you in its silver liquid drops. Embrace the renewal it brings, as it cleanses and rejuvenates your soul.

Aurora and the Zodiac, Jacob van der Schley, 1725 - 1779
In the universe, Aurora floats above the Earth. Part of the zodiac can be seen in the background; the scales, scorpion and the archer.

Go, go, and seek the beauty of life!

Enter the woods, let your feet tread on soft moss and pine needles. Breathe in the scent of cedar and let it fill your lungs. Let the rustling leaves be a symphony for your ears. Let the damp earth and the rustling leaves remind you of the fleetingness of life.

Emerge from the forest reborn, taller than the tallest tree.

Seek out the beauty of life amidst the chaos and mundanity of the world.

But do not hurry.

"Nature does not hurry, yet everything is accomplished."
Lao Tzu

In dwelling upon the beauty of Nature, you'll find the fount of strength that will sustain you until your last breath.

Because,

"Just living is not enough. One must have sunshine, freedom, and a little flower."

Hans Christian Andersen.

Maxim

7.48

"This is beautiful in Plato. And when the talk is about mankind, one should
TAKE A BIRD'S-EYE VIEW OF EARTHLY THINGS
and look down on them as if from somewhere above."

Marcus Aurelius. Meditations 7.48

Imagine you shed your skin like a snake, slip from your bones, and soar into the atmosphere, soaring higher and higher until you are nothing but a speck, a grain of sand on the earth below.

You see yourself, your house a dollhouse on a carpet of green, your neighborhood a jigsaw puzzle of streets and houses.

You see the ants below, scurrying to and fro, the human anthill in all its frantic businesses. Traffic jams, like rivers of metal and glass, flow through the streets of your sprawling town.

And further up, you see your country, a patchwork quilt of fields and forests, mountains and deserts.

The ocean, a vast and endless blue, stretches out before you like a dream.

And then the whole planet, a fragile marble hurtling through the void, is yours to behold.

You are a tiny pinprick of light among a sea of stars.

Like an astronaut, wide-eyed and weightless, you peer out at a pale blue dot, our fragile orb suspended in the vast and merciless expanse of space. All the wars and divisions that plague our planet disappear in that one, transcendent moment, revealing nothing but the delicate beauty of our tiny home, a precious and fleeting spark in the boundless darkness, a dead, cold, ruthless, and indifferent void.

"Look again at that dot. That's here. That's home. That's us. On it everyone you love, everyone you know, everyone you ever heard of, every human being who ever was, lived out their lives. The aggregate of our joy and suffering, thousands of confident religions, ideologies, and economic doctrines, every hunter and forager, every hero and coward, every creator and destroyer of civilization, every king and peasant, every young couple in love, every mother and father, hopeful child, inventor and explorer, every teacher of morals, every corrupt politician, every "superstar," every "supreme leader," every saint and sinner in the history of our species lived there—on a mote of dust suspended in a sunbeam.

The Earth is a very small stage in a vast cosmic arena. Think of the rivers of blood spilled by all those generals and emperors so that, in glory and triumph, they could become the momentary masters of a fraction of a dot. Think of the endless cruelties visited by the inhabitants of one corner of this pixel on the scarcely distinguishable inhabitants of some other corner, how frequent their misunderstandings, how eager they are to kill one another, how fervent their hatreds.

Our posturings, our imagined self-importance, the delusion that we have some privileged position in the Universe, are challenged by this point of pale light. Our planet is a lonely speck in the great enveloping cosmic dark."

These are the thoughts of Carl Sagan commenting on a famous Pale Blue Dot photograph of planet Earth taken by the Voyager 1 space probe from an unthinkable 3.7 billion miles away. In the photograph, Earth is a speck, a pinpoint in a sea of black, less than a pixel in size, lost among the stars and the bands of light reflecting off the lens.

Now the descent begins, falling back to Earth, through the stratosphere, back through the clouds.

Your continent, country, and city grow larger with each passing moment. All of humanity's joys and sorrows play out in that small space. Babies born, with tiny fists clenched, seeds planted, dogs rolling over, tongues lolling. People meeting, hearts colliding and breaking, memories fading like photographs left in the sun.

Life, ever-changing, forever moving forward.

And now your street, your home, your room, your chair, and finally, you, sitting, back where it all began.

You slip back into your anxious skin, the familiar ache of worry etched deep within. It's still important, still unresolved, but now it doesn't feel so crucial. Your current troubles may loom large, but in the grand scheme of things, they are small, just a drop in the ocean of your whole life.

And your life is only one of billions on this spinning rock, each with its own struggles, each with its own worth. No matter what happens in any of these lives, the stars will still shine, bright and beautiful, unconcerned and unaware of the newest wrinkles etched in your skin...

This is a pure experience of perspective and context.

The weight of our problems can suffocate us, blinding us to the larger context in which they exist. Like waves crashing against the shore, our problems may seem insurmountable, but in the vastness of the ocean, context sets them in its calm. We tread the water's edge, blinded by the tide's roar, while the answers we seek lie in the depths below.

At some point in high school, you probably felt the weight of the ocean on your shoulders. Exam season was a trial by water, and your entire future seemed to hinge on a single test. But now, in retrospect, that moment is but a blip on the radar of your life. It barely registers as an important event.

Why is it that, in the moment, we cannot see the insignificance of our worries?

Worry is a tide that swells, consuming the horizon until all that remains is a narrow, fixated view of our current troubles. It is the feeling of being pulled under by the weight of our thoughts, the ocean of our mind filling with the endless ebb and flow of worry.

But as the tide recedes, you are left with the vast expanse of the ocean, where that one moment of stress is but a ripple on the surface. In retrospect, you see the insignificance of that exam in the grand scheme of your life.

Usually, it is only in retrospect that we understand the power of perspective. The ability to step back and witness the ebb and flow of life, to see that even the most tumultuous of waves eventually subside.

So you must remember to take a deep breath, step back, and see the beauty and majesty of the ocean of life that surrounds you. And to let go of the worries that threaten to consume you. For in the end, they will be but a distant memory, lost in the depths.

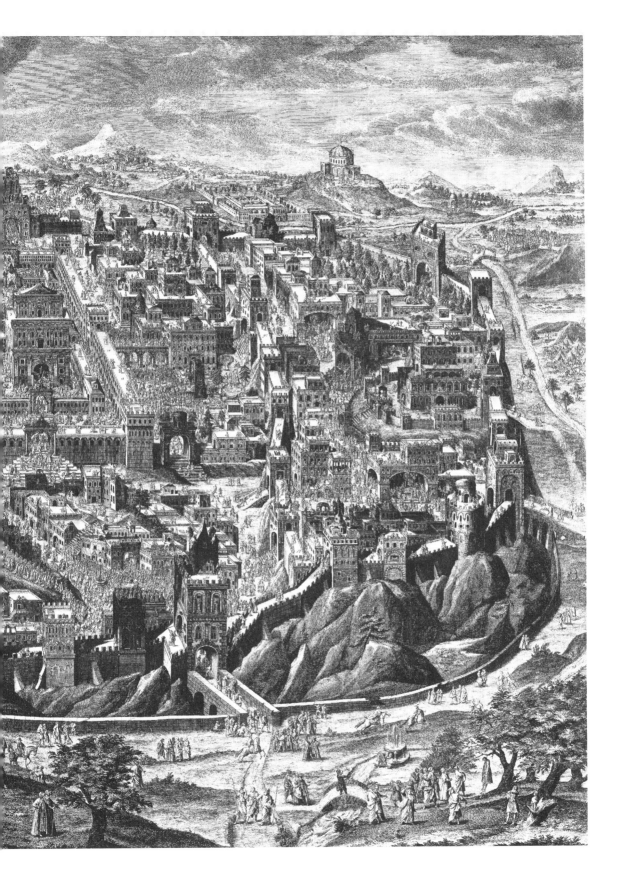

Bird's-eye view map of Jerusalem depicting Christ's Passion, Johann Daniel Herz (I), c. 1735

This is what *"The bird's-eye view"* is about. The Universe, vast and indifferent, makes your troubles small and fleeting. They shrink to nothingness.

This is where perspective is found, where the bigger picture comes into focus and the illusions of wealth, looks, and status fade away like stars in the daylight.

For when you're too close, you lose sight of the edges, the before and after, and the world beyond your worries. So you must zoom out, to encompass all that exists outside of your petty concerns.

The Stoics say it's all about how you look at things, that it's not the people, things, or events that break us but the way we choose to see them. We hold the power to turn obstacles into opportunities *(Maxim 8.32, "Arrange life act by act")*, to view insults as jokes, and to laugh at them instead of letting them sting.

We hold the power to detach ourselves from our worries. It's all about a shift in perspective. It's like standing on a mountaintop and seeing the world from far away—suddenly our troubles seem small and insignificant in the grand scheme of things.

"You can eliminate so much of what weighs upon you as rooted wholly in your judgments. And a wide expanse will straightway open before you for embracing in your thought the whole Universe, for grasping everlasting Time, for realizing how quickly any part of any thing does change, how truly short it is from birth to dissolution."
Meditations, 9.32

Next time you're feeling down, take *"The bird's-eye view."*

Imagine, across the miles and centuries, an emperor of Rome, in all his power and might, finds the same relief in contemplating the same boundless expanse that you do.

Maxim
7.56

"As one who's dead,
whose life till now lived and gone,
resolve to live the rest in sweet accord with nature."

Marcus Aurelius. Meditations 7.56

Imagine you awake to the reality that death has claimed you in the night. You've passed in your dreams. The world moves on.

Again. You are no more. Gone, vanished.

Did you live a life of truth and beauty? Have you even lived?

Think well.

Each dawn is a chance to grasp at what remains, to make the most of your fragile existence. A chance to live, love, laugh. Make life.

Life's secret heart is this:

"Die before death.

Embrace the fleetingness."

You must die before you truly live, and in this realization, hold dear your own impermanence.

Memento Mori.

Death strips bare all that is not you, i.e. not true. Imagining it is key to living well.

"Memento Mori" is Latin for *"Remember you must die."* It is said to have begun in ancient Rome. After a military victory, a companion or slave would walk behind the victorious general during his military triumph. The slave's job was to whisper from time to time in the general's ear of the fleeting nature of mortal existence.

"Respice post te. Hominem te esse memento. Memento mori!" (*Look behind. Remember, you are mortal. Remember, you must die!*). The general would remember that victory, like life, is transient and not to be assumed as a given.

Memento mori is a reminder of life's fleeting nature and the certainty of death. A call to live with purpose, to let go of worries beyond our control.

Though it may sound strange, death is the key to a life full of joy. The contemplation of death should bring not sorrow but a heightened range—of truly living, in the present. In the face of death, we are reminded to live, to truly appreciate being here, right now (Maxim 12.3, *"Learn to live in what alone is life—the present"*). Embrace death, dwell on it, make it your constant companion. In the shadow of death, life blooms with vibrant hues. The thought of death brings depth to the days. No more carelessly spent hours, a newfound appreciation for life's simple pleasures, pure and free. Each moment to be savored, no longer taken for granted, honored and favored like the sun setting on the horizon. Choosing purpose, no aimless treading—only important things, they will last.

Death looming is a reminder to live fully, be bold. Let go of the meaningless and focus on what's true. In this world, in this life, who you want to be, and with whom (Maxim 8.2, *"At any action ask, "What does it hold for me?"*).

Your heart is never the same. It flicks like a flame in constant motion. You cannot know how long it will continue to flow. All you can know for sure is how you choose to live in this moment, now. Marcus speaks true, live each day as if it were your last. *"Act, speak, and think as one, who may depart from life at once"* (Maxim 2.11).

Living as if each day is the last is not about grand gestures or feats. It's to remind yourself, in every way, of life's fleeting span. That you are mortal, with a limited role. Some day, like a wave, you'll crash upon the shore. It's not about a major shift in what you do, but how you do it; it's a mindset shift.

Life's a fragile thing, a fleeting breeze. Be grateful for each breath. Connect with those around you. Make each moment count. In the end, it's the human connection that you will miss.

Vanitas still life with skull, Jan Saenredam, after Abraham Bloemaert, 1575 - 1607

Don't let the hours slip away, lost in trifles and nonsense. With purpose and resolve, choose to live, not just be, in harmony with Nature,

"and fear not that life will ever stop, but that you never start at all to live with Nature's order."
Meditations, 12.1

Living in accord with Nature is to honor the human nature within you—to be just, wise, balanced, and brave. In every breath and step you take *(Maxim 2.5, "Resolve to do the work in hand with scrupulous and unaffected dignity")*.

Some people get a *Memento Mori* tattoo, forever inked as a reminder upon their flesh. In fact, you can simply look at your palms—hand lines make the letter *M* on each of them: M. M., Memento Mori.

Look often, and remember: *You must die.*

The most effective *Memento Mori* practice is perhaps *active imagining.*

Imagine you're a corpse. In a wooden box, beneath the earth. Lying there, in the stillness of death. Imagine the worms feasting, the bones bleaching white. And then a pile of dust, all that's left of those.

Now, look at your life from that final viewpoint. Do your problems not seem like a bunch of nonsense, a mere mirage?

Think of moments unnoticed, unlived...

From the vantage point of death, each fleeting moment holds weight.

So take a moment each day to view the world through death's eyes. You'll come alive in ways you never ever thought were possible. Feeling the weight of every breath, every beat of your heart. Life as a symphony of sense and sound.

In fine:
Imagine you are dead, and take that emptiness,
and make it yours,
and live it properly.

Maxim 7.57

"To love what happens only, woven in the web.

For no fits better, ever."

Marcus Aurelius. Meditations 7.57

What do you want? Wealth, fame, success? The promise of forever love, eternal friendship?

Marcus desired none of these. Only what his control could bring him. And trusted in the power above *(Maxim 9.40, "Turn your prayers, see what comes.")*. For him, all that happened had a purpose.

And life is not a pony farm, of course. Shit happens, that's the norm. Instead of resenting or longing for what's not there, work with the cards you've been dealt… A familiar tune, no denying. *"If life gives you lemons, make lemonade,"* as the meme goes. But Marcus goes even further. He transcends the mere acceptance of fate. Rather than just surviving circumstances, *love* them.

The goal is not a whisper but a scream.

"This is just what I need! Damn, I'm glad it's here!" instead of just *"Ok, I'll take it on."* Not a vague acceptance, but a burning desire. And with that comes a feeling of euphoria, a sense of fulfillment.

"If it's here, it was meant to be."

"I'll make the most of it."

"I am filled with joy, I love it all."

If it must happen, then *Amor fati!*

Amor fati means "love your fate," which is in fact your life. While the ancient Stoics had the idea of *Amor fati*, the words came later. Friedrich Nietzsche coined the phrase *"Amor fati,"* as *"a passionate acceptance of everything that happens in one's life."* Here is the essence of *Amor fati* from Nietzsche's Ecce Homo:

> *"I want to learn more and more to see as beautiful what is necessary in things; then I shall be one of those who make things beautiful. Amor fati: let that be my love henceforth!"*

Amor fati is to see the beauty of fate and fall in love with it. It is the mindset of taking it all in, the sweet and the harsh. Embracing the hardships and struggles as necessary parts of life, whether we like it or not.

Sure, Lady Fate's allure may not be so sweet at a glance. So why would you actually do A*mor fati?*

Well, we all want to be happy. Moreover, we want to be happy immediately. For there is no other happiness, just as there is no other time but now. Happiness is when the soul feels good right here, right now.

But no matter how hard you try, you can't achieve "happiness" exclusively on your own. Just because the rest of the world plays a role in each of our moments and we have little control over the world. That's why there is only one path to happiness:

Accept deep in your heart what is happening right now in the world. Love it as if you've been dreaming about it your whole life.

Just dissolve into the moment.

Suddenly, a feeling comes that everything unfolds with purpose. And once you stop struggling with it, you'll realize how you could use it to your ends. It's up to you alone to shape that purpose into something bright and active.

The loving acceptance of fate is power so grand and vast that it is even hard to grasp. It brings gratitude and peace, stability and resilience, well-being, and serenity.

It brings the understanding that change is the very essence of the Universe. That without it, we would not be, our relationships would not be, we would not laugh or cry, love or create, grow or live. We just would not know life (Maxim 4.36, *"Ever behold as all things come to be through change"*).

It's safe to say that nothing ever stays the same. And even if it could, why would we want it to?

"Afraid of change? And what can happen without change? What, after all, is dearer or more proper to Nature? Can you even take a warm bath if the wood does not turn into heat? Or can you be nourished unless food is transformed? Can anything useful be done without change? Don't you see, then, that for you to be transformed is something equal and equally indispensable to Nature?"
Meditations, 7.18

Change is necessary, whether it's good or bad, or whether it brings pleasure, pain, or loss. Resisting, hating, or complaining about fate is resisting the very conditions that have allowed you to be.

Everything that has happened before you, every change and transformation, has led to this very moment, where you may read these lines. In that sense, you have no other choice but *Amor fati*.

"For nothing is self-sufficient, neither in us ourselves nor in things; and if our soul has trembled with happiness and sounded like a harp string just once, all eternity was needed to produce this one event—and in this single moment of affirmation all eternity was called good, redeemed, justified, and affirmed."
Friedrich Nietzsche, The Will to Power.

That's why you shouldn't reject anything. Next time the rain falls, don't curse the weather. Embrace the feeling, find the music in the drops, discover joy.

Loving tough times reveals even more of life to love. Wish not for time to fly, for you are wishing away moments that will never return. Moments that could have been filled with love.

Sure, it's harder to love fate when it's harsh. Still, that's fate, that's life. *"This too shall pass."* But if you ready your heart with *Amor fati* response beforehand, you shall not be caught unawares by destiny's blows. For *Amor fati* is not a mere surrender to the winds of fate, a mere *"Soaring above the volatile times."* Loving acceptance spares you the wasted hours of cursing fate and lamenting luck. You don't waste your time wishing for things to be different. You focus on action, on what you can do. You assess the situation, make a plan, and take control of what you can.

Here is an *Amor fati* equation for *"Volatile times"*:

Amor fati = Define + Accept + Act

1) Define

The first thing we must remember is that whether an event is good or bad is a matter of perspective. One man's tragedy is another man's triumph. Objectivity is key. Before forming an opinion, aim to see the event for what it is, objectively *(Maxim 6.13, "Strip the facts bare")*.

2) Accept

The issue is now clearly defined. You need to accept it.

A man taking on the stance of *Amor fati* is a man who does not only accept all that comes, but embraces it with open arms and a fiery heart of gratitude, turning even the darkest moments into something of worth. Adversity may be harsh, but it is never the worst. Think of what it may have spared you from.

Beauty can be found in every new experience, because it is at least a learning opportunity and a chance for something equally unexpected and positive to happen as a result.

"I am in tune with all that is your harmony, O World. What comes in time to you is neither early nor too late to me! All things are fruits to me, what your, O Nature, seasons bring. All things from you, all things in you, back unto you, all things!"

Meditations, 4.23

This is how Marcus Aurelius expressed his will to embrace what is given.

Craft your own acceptance mantra to affirm in the midst of adversity, or use the above.

3) Act

Weigh all options within your power to solve the problem, which already has an objective definition.

Choose the top choice and split it into smaller steps.

Act on the initial step *(Maxim 8.32, "Arrange life act by act")*.

Remember, when things get tough, there's always a way to find a silver lining. It's all about a shift in perspective, seeing the bad as a chance for good.

To move from mere accepting to *loving* your fate, you must choose to see your struggles as challenges and chances. Challenges reveal your strength. Like a match to a fire, obstacles, and hardships become the spark for your growth.

"A blaze of fire turns everything you throw at it into flame and light."
Meditations, 10.31

Negative thoughts can consume your mind like a raging fire. But positive thoughts can harness that fire, using it as fuel to drive you forward.

Now back to the lemons. They pucker the mouth, leaving a bitter tang on the tongue. But squeeze them, mix in sugar and water, and suddenly they sing a different song. This is the alchemy of life.

When the bitter moments come, instead of dwelling on the harshness, reframe your view of it. And with a little bit of work and a little bit of tweak, you can turn the bitter into something truly unique.

Be creative.

When life gives you lemons, order tequila and enjoy.

Let your life be full of happiness, tequila, lemonade, or at least so many lemons that you can open your own lemon shop.

Cheers!

The Inconstancy of Fortune; It is Easy to Dance when Fortune Plays for You, Dirck Volckertsz. after Maarten van Heemskerck, after c. 1560 - before 1590

Lady Fortuna stands with a man holding a laurel wreath on a seesaw. Her side slopes down. The seesaw is placed on the back of a dolphin. A rainbow in the background.

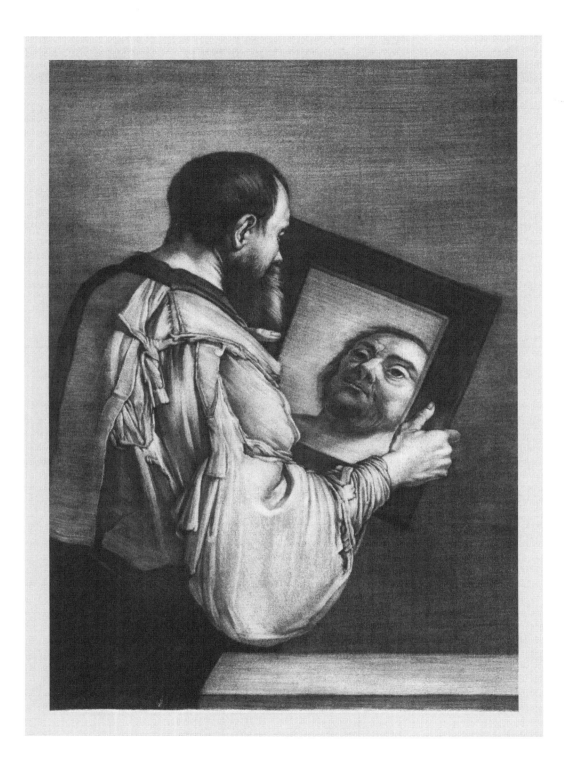

Socrates Looking at Himself in a Mirror, Bernard Vaillant, after Jusepe de Ribera, 1672

Maxim

7.59

"Dig within. Within you is the fount of Good,
ever you dig, and it will ever flow."

Marcus Aurelius, Meditations 7.59

This spring of good may be equated with the inner spirit or the inner daimon. The Stoics believed that *eudaimonia*, or being good *(eú)* with one's own inner *daímōn* (tutelary deity), was the ultimate purpose of life.

Daimōn was translated into English as "demon." But the two are not synonymous in essence. Marcus Aurelius often employs "daimon" as a substitute for "reason." Yet, in other moments, he draws a distinct line between the two, revealing the daimon as a spark of the divine, a fragment of the greater deity—be it Nature or Zeus—nestled within each of us.

The Stoic Daimon is a *"God lodged within your chest."* The Egyptians had a similar concept they called "Ba." According to Aristotle, the daemon represents the highest and wisest part of a person and is where their true potential lies.

The challenge with this entity is that it's always lurking, unseen. We all carry a daemon inside—a voice that whispers in the dark, a flicker of intuition. But society's chains can make it hard for the daemon to spread its wings and fly. Your daemon, caged by upbringing's bars and stress's chains, is a slave to anxiety, schedules, and a constant preoccupation. To free your daemon and hear the voice, you must know yourself. Unearth your own roots.

So dig within.

When does the digging begin?

In those moments when the soul feels weak. When your moral compass starts to waver. When your actions are going to betray your values. When you find yourself tempted by the allure of wrongs. When the truth of who you are is obscured by lies.

Where to begin the digging?

Within oneself. As with all things of value. For the good, like all true treasures worth possessing, is already within you. Always. Often buried beneath the layers of fear and doubt. Dig where the roots of compassion and integrity run deep, where the seeds of hope and goodness have been planted and are waiting to bloom.

Dig where your daemon is.

Your inner demon, your divine companion, plays three roles:

- *The guardian of destiny*—the force that guides you towards your truest self, revealing the innate temperament, inclinations, and tendencies that shape your life, even your personal struggles.
- *The inner judge*—the divine presence that watches over your actions, words, and thoughts, holding you accountable for the good and the bad you do, and granting you the grace to learn from your mistakes and become better.
- *The guardian angel*—the soft whisper of guidance that speaks to you in moments of uncertainty, the unerring voice of wisdom that steers you towards the path of virtue, the dream that holds the answer to your deepest struggles.

In short, your daimon sets the course of fate, shapes the moral fabric of your being, and guides you towards good.

To listen to the whispers of that inner voice is to establish a symphony with your innate divinity, thus summoning the "good demon", or *"eu-daimonia"*—happiness in its finest form, a balance of goodness and truth.

According to the Stoics, the fire of the inner daimon, this "spark divine", fuels us to become the best version of ourselves. Like a seed, it's been planted within us, waiting to be nurtured and grown into something magnificent. It's our nature to complete the journey, to bring our potential to life.

The Stoics had a term for that elusive state we all strive for: *arete*. It translates to "virtue" or "excellence," but it reaches deeper. It's about aligning with the best version of yourself in every breath.

To become one with the divine voice within, to live in alignment with your truest self, is to ascend towards the highest version of yourself, to close the chasm between what you could ultimately be and what you currently are.

How is this even possible?

You can use either a logical or a more creative, intuitive approach (or both).

In Greek tradition, the daemon was also the highest source of creativity and inspiration. Metaphorically speaking, being in touch with your daemon is to transform your monsters into muses.

There is a belief that madness burns beneath the surface of art and that the broken are the ones who create. We see it in the tortured brilliance of Virginia Woolf and Vincent van Gogh, their art born from the ashes of their despair. But what about the beauty that comes from a mind at ease? True creativity is nourished by the daemon that brings us happiness, *eudaimonia*. It is our well-being that enables us to tap into our potential and live a life of meaning and fulfillment.

The mind at ease, the heart at rest—that's where the magic happens and the most fruitful art comes from. But the journey to that place is one of exploration.

Maybe you find insights in the wild, in music, or in the company of others. Some prefer seclusion, while others immerse themselves in the beauty of art, and the creativity flows. To nurture intuition, you must tend to it like a delicate flower. Meditations, yoga, the solace of a walk, the quiet of contemplation, or the scribbles of a journal can all allow the voice within to speak its truth.

"When you close your doors and create darkness within, remember never to say that you are on your own, for in fact, you're not alone, because God is within you, and your guardian spirit too."
Epictetus, Discourses (I, 14)

Always remain open to new ways and practices that allow you to uncover the wisdom and hear the message of your inner guide. As Rudyard Kipling said, *"When your demon is in charge, do not try to think consciously. Just let go, wait and obey."* This inner voice is a thread woven into the fabric of your being. It is only when you've peeled away the layers of pretense and facade, when you truly know who you are, that you can hear its call.

So, if you take a logical approach—choose what you will, and dig, dig deep as Marcus did, not with a shovel but with the sharpened edge of your mind. Ask the questions and answer them honestly.

Why do I do it? What's the point? What's the worth in it? *(Maxim 8.2, "At any action ask, "What does it hold for me?").* The reason here is the compass, guiding you through the chaos. Hold tight to your values while navigating towards harmony.

Reason and values, a dance in harmony...

Easier said than done, for sure. Just always strive to distinguish right from wrong and claim power over what is within your grasp.

Whichever path—intuition or logic—you take, keep the Good flowing, and keep digging—anytime, anyplace—where the fountain falters.

Zeno said happiness is like a river that flows with ease. A current that carries you smoothly through the landscape of your life.

But what is this smoothness?

It's the alignment of your actions, moment to moment, with your truest self and the daemon within.

Dig within and listen to your inner deity.
And let the eu-daimon possess you!

Maxim

8.2

"At any action ask yourself, "What does it hold for me? Shall I repent of it? A little while and I am dead, all these things are gone."

Marcus Aurelius. Meditations 8.2

Life is short, a fleeting thing. A feather dancing in the breeze of time. And yet, a lot of people waste it chasing after things that hold no weight, getting lost in aimless wanderings, indulging in mindless distractions that only serve to numb the passage of time.

Killing the time. Regretting it all.

And then again, from day to day, year to year, people are killing the time as best as they can. Hearts cold and empty, eyes without cheer, tears dried to salt on their cheeks. Regret is like a blade that carves the days. And the wind outside is howling its fury, a reminder of all they've failed to hold…

The sad truth is that you cannot waste time. You can only waste yourself. By forgetting that you're mortal. By living as if you will never expire. Until you realize that you will. And that's when you wish for a second chance—that you'd changed everything and started earlier to actually live.

But life does not get better by chance. It gets better by change.

So ask yourself, Will you be the one standing at the edge, unprepared for what's to come? Your heart heavy with missed opportunities? If you find yourself acting as though death will never come for you, remind yourself that you are mortal, embrace your transience *(Maxim 7.56, "As one who's dead, resolve to live the rest in sweet accord with nature")*, and cease to waste your days on meaningless pursuits. You have to eliminate the nonessential *(Maxim 4.24, "Do few things — if you seek cheerful calm")*, and start to purposefully fill up your days with things of value and meaning.

In short, stop doing the things that do not serve your deeper life goals and do only what is necessary for achieving them.

Pause before diving headfirst into any activity, and consider if it will truly benefit you as a just and reasonable human being. The key here is to keep an eye out for the early warning signs. The habits that consume us, whether in the currents of social media or the depths of grievance over the past, are mere moments in the flow of time. To change their course, we must first become aware of them, which often occurs only when we are already caught in their tide.

Practice and hone your reflexes. Soon you'll be able to catch yourself in the act, at the beginning, or before the descent even starts. The secret to it is constant self-examination, noting the subtle movements of muscle and mind as you walk through the day.

The Stoics practiced an awareness technique they named *prosoche*, which literally means "attention" *(Maxim 12.3, "Learn to live in what alone is life—the present")*. For example, you feel the shift in breath, the flutter of your pulse, or the tightness in your throat as you start to drown in the work and consider drowning in the news feed as a relief. Then, before the action, stop and

"ask yourself, Is this a necessary one?"
Meditations, 4.24

It is a way of revealing the consequences of your choice. You inquire, "Is it vital?" Does it nurture you, or does it harm you and those around you? Does it bring you closer to your goals or push you further away? In the end, it all comes down to one question: *Is it worth it?*

Questions like these steer you towards the purpose of whatever you're doing. Where is it all heading? What does it hold for you? Look closer at the end result of your actions. Is it truly leading you to where you wish to be on your journey?

Only when you've learned to see the early signs of the behavior you want to quit and questioned its value or impacts, can you break free from the chains of habits that lead to nowhere.

But with that newfound freedom comes a question: "What to do with all the time that has been freed?"

Keeping in mind that *good is not in goods*, ask yourself, What holds the most value in your life?

It is the ultimate question. Beneath the clutter of wants and possessions lies a deeper yearning, but can you bring it to the surface and give it shape with language? Can you name your core values and truly understand them in your heart? What do you hold dear in the silence of the night? When all is stripped away, what remains in sight?

The answer may not come easy, but it's worth the search. Look within, unearth the things that matter most, that give life worth.

"To what use am I putting my soul right now? Ask yourself this question all the time. Examine closely: What is it now in that part of me known as the guiding Part? And whose soul is mine at the moment? — isn't that of a child? Or a boy? Or a woman? A tyrant? A cattle? A beast?"

Meditations, 5.11

What things do you really wish to do in life, what love to try?

What will you fight for? What do you burn for? What's the fire in your bones?

Where do you run? Quo vadis? What will they say of you when you're gone?

As an alternative imagine the inscription that will mark your final resting place—a mere sentence or two to sum up all you've done, the meaning of your life. What do you wish it to say?

Remember, your values are not just "goals" to "achieve," but rather paths to be followed in life, constantly etched in the lines of your actions. The principles that shape the way you live, ever-unfolding.

"Being a good friend," it really never ends.

"Justice in action" cannot be grasped like a prize, held tightly in clenched fists. It's a pilgrimage to be embarked upon. It's the ongoing journey of aligning your actions with your deepest beliefs, never arriving at a destination, but forever moving forward.

"Integrity" is not a possession, but a practice. You do not arrive at it, but rather, you live it. It is a way of being, a constant cultivation, a forever blooming. It is not an end, but a means—a path to walk, hand in hand with your truest self.

Values are like stars in the chaos, guiding you northward with a steady glow. They are not a destination, but a compass pointing you ever forward, always true. Not a finish line, but a path to tread. A map to navigate.

When they shine brightly in your mind, your deepest values give form to the backbone of your existence, shaping the philosophy you live by. You must return to them again and again, checking each day: "How much of my life did I live in harmony with values of mine?"

In the end, your values become your destiny. Your value is your values' total.

So, let your values be your breath,
The pulse that beats within your chest.
Speak them, wear them,
Align your truth with them.
And use your time so wisely,
That in the very end, when you look back,
A smile will greet you,
Not the echo of regret.

Allegory of Time and Fortune, Monogrammist AC (16th century)
Two naked young men at sea in a shell with a sail. One is holding an hourglass. A helmet and the shield lie in the shell.

Fable of the Tortoise and the Hare, Aegidius Sadeler (II), 1608–1679
A tortoise and a hare are standing by a fox. In the background is a city. The fable describes the running race between the tortoise and the hare. The hare is much faster than the tortoise. He thinks he has already won and goes to rest before he has reached the finish line. The turtle moves slowly but has perseverance and crosses the line first. The moral of the story is that perseverance is more important than talent.

Maxim

8.32

"You must arrange life act by act, *and be content if each act as it may fulfill its end; no one can hinder you in that."*

Marcus Aurelius. Meditations 8.32

"Divide et impera."
("Divide and rule," Lat.)

The words of a conqueror (usually associated with the military genius of Julius Caesar but belonging, most probably, to Phillip II of Macedon) echo through the ages, a bitter maxim for empire-builders who seek to expand their reach by turning their foes against each other. A strategy as ancient as empires, a way to rise above by making others fall.

And at the same time, the maxim "Divide and rule" (not the original idea, though!) could be successfully applied to every aspect of life, being just a strategy for those who seek to claim in every endeavor.

The weight of a vast and intricate problem can be suffocating, leaving one to question their trust in their own capacity for resolution. "Divide and rule" is a way of breaking down the impossible into something that can be conquered, a game of slicing through the heart of the big challenge and breaking it down into manageable chunks until all that remains are the easily solvable pieces. Then, you must hone in on a single, solitary issue and let nothing else distract you until it's been unraveled. Only proceeding to the next, once the present has been resolved.

Thus, brick by boring brick, the wall is being built.

Just show up, day after day, despite the trials that may come, with steadfast grace and a heart full of patience, until the finish line is crossed.

"It's not that I'm so smart, it's just that I stay with problems longer," said **Albert Einstein.** *"I think and think for months and years. Ninety-nine times, the conclusion is false. The hundredth time, I am right."*

This is Persistence.

We are made to persist. That's how we discover who we are.

But what about external obstacles?

"If outer impediments arise? — None to actions that are just, moderate, and wise. The other sorts of action may be hindered. But if you welcome the impediment and smartly adapt what was granted, an alternative straightway will take its place that fits the course you are arranging, act by act."
Meditations, 8.32

Sometimes obstacles may loom like monsters, unyielding in their might. But what choice do you have?

Embrace the obstacles; make them a part of the game. Keep moving forward, keep doing the work, another path will unfold, Marcus writes. Just trust the journey, focus on the process, and hold tight to what you can control.

A famous archer metaphor illustrates the fundamental stoic principle of controlling the controllable while releasing all else to fate's grace.

The archer tugs the string, one eye closed to sharpen his aim. They hold their breath, let the arrow loose, and watch as it soars towards its destination. But there's no telling what could happen now—the wind could shift, a bird could dart across its path. All they can do is wait and watch as fate and circumstance determine its final destination.

The point is, the archer can draw the bowstring taut, aim with precision, doing all their best up to the moment they release the arrow, but whether it finds its mark is not in their hands anymore.

So too in life, we set our sights, take action, and like the archer, we too can only do so much. The rest is a dance between what we want and what the world allows.

We can choose our steps, but the rhythm is set by the world. The Stoics teach us to let go of the end and embrace the journey, to find beauty in, and to focus on the process (which is under our control) instead of the desired outcome (which is not).

To strive is to succeed. The true measure of success will then be the fullness of your effort, your willingness to pour every last drop of your being into the task at hand.

With such a marker of success, give all you have, and you will never fail. Because your efforts are within your power. In this way, peace will always be within your reach, you'll always walk with grace and confidence, unshakable, unbreakable, at ease with any outcome fate brings.

The archer, stoic in his aim,
Finds solace not in the bullseye,
But in the steady pull of the bowstring,
The release of an arrow into the unknown.
Victory and defeat are mere whispers,
In the deafening silence of the process.
He succeeds, not in the end result,
But in the motion of becoming.

So, build your life, act by act, brick by brick. Each step a heartbeat, each obstacle a lesson. Learn to weather the storms, and in the process, find peace and growth *(Maxim 7.57, "Love what happens only, woven in the web")*.

Keep going. And always remember enjoying the journey.

Maxim

8.36

"Let not the visions from the entirety of life dismay you.

Don't dwell on the kinds and scale of the pains that may come. Of any situation in the present, ask: What is there I cannot endure? Or bear? The answer will put you to shame!"

Marcus Aurelius. Meditations 8.36

Have you ever come face to face with a task so daunting that it felt like scaling a mountain with bare hands? The journey ahead looms, a path lined with obstacles and insurmountable goals. A chasm of effort and uncertainty. The sacrifice required to reach the summit seems almost too much to bear.

In those moments, the weight of the world presses down, threatening to crush you under its enormity, and you feel the despair of futility creeping in, tempting you to give up.

As you feel the world closing in, pause for a moment and remember Marcus Aurelius. Emperor of the Great Empire, battling plagues, wars, and natural disasters. His mind plagued by illness, his treasury drained by the financial crisis. Imagine the weight of it all, the constant barrage of loss and sorrow. Imagine the pain of watching as five of his own children were taken from him, one by one. And yet, amidst it all, he found a way to rise above. He found a way to hold on to his integrity, his wisdom, and his compassion.

Strength (Fortitudo), Jacob Matham, after Hendrick Goltzius, 1587 - 1637
The ***Strength*** personified; a female figure with a broken column over her right shoulder. In the background is a hilly landscape.

Here is what the historian Cassius Dio wrote of Marcus' calm perseverance:

"[Marcus Aurelius] did not meet with the good fortune that he deserved, for he was not strong in body and was involved in a multitude of troubles throughout practically his entire reign. But for my part, I admire him all the more for this very reason, that amid unusual and extraordinary difficulties he both survived himself and preserved the empire."

How did he make it through? How did he bear the weight of all those fears and doubts?

Marcus refused to let the "visions of an overwhelming future" suffocate him, and didn't let the thoughts of the "scale of the pains that may come" consume him. Instead, he anchored himself firmly in the present, aiming to live in each moment as it came. He methodically broke down his struggles into manageable pieces, set deadlines for himself, and moved forward, one step at a time, leaving everything else to drift away.

First, *"anchoring the present."*

Why is this of such importance that Marcus mentions "the present" and "the present moment" more than twenty times in Meditations?

Our war is this: we are torn by our thoughts, like a ship caught in a storm. We're tossed between memories of what was and worries of what will be. And all the while, the present slips through our fingers like sand. We forget to live in the present, to feel the weight of the moment. This is the heart of our overwhelming.

But in this daily tempest, we must remember that the past and the future are beyond our control. It is only in the present that our power lies. For the past is but a shipwreck washed ashore, irretrievably lost. And the future is but a horizon waiting to be reached with each step we take. It can only be influenced by our actions in the here and now.

The only thing we truly possess is this moment, this breath, this beating heart, and that is where we must anchor ourselves *(Maxim 12.3, "Learn to live in what alone is life—the present")*. All the force we possess condenses into this one instant—now. The choices we make are the only true power we have.

The thoughts we choose, the actions we take, are the only things that belong to us. Only now, only here, do we truly have control. Only in this fleeting moment do we truly reign. Thus, when overwhelmed with the issue at hand, ask:

"What is there I cannot endure? Or bear?"

Then, anchored in the present, focus on what is tangible and true and move on, one small step at a time, until the unthinkable becomes possible. Hone in on a single step, striving for excellence in its completion. Divide an overwhelming "chaos" into a sequence of achievable tasks. Attend to each one without allowing the bigger picture or upcoming tasks to disrupt or dishearten you *(Maxim 8.32, "Arrange life act by act")*.

Attack each task with unwavering determination and assurance.

"At any moment, firmly, as a Roman and a man, resolve to do the work in hand, with scrupulous and unaffected dignity, with justice, tenderly, and willingly;"

Maxim 2.5

So when the tide rises and the waves crash against you, do not falter in the face of their enormity. Remember Marcus' wisdom:

Narrow the view,
To not be drowned by troubles dire,
Perseverance, the key,
One step at a time, move higher.

Fable of the Young Man and the Swallow, Aegidius Sadeler (II), after Marcus Gheeraerts (I), 1608

A young man, dressed only in a shirt, finds a dead swallow in his way. The fable describes how a young freeloader sees a swallow in the air. He considers the swallow a harbinger of warm weather and therefore sells his clothes, except for his shirt. Later, the youth finds the swallow frozen to death by the roadside. The moral of the story teaches that one should not judge lightly and should think before acting.

Maxim

8.49

"Tell yourself no more than what's declared by the first impressions."

Marcus Aurelius. Meditations 8.49

That is not easy at all. You give meaning and interpretation to everything. It's like a reflex, a knee-jerk reaction. A natural compulsion to read into the world, to gain senses from the chaos. An instinct as innate as the breath you take.

It ain't no simple task to give it up, but it is possible.

Before you may treat impressions as mere hypotheses, your eye must first discern them in their fleeting poses. This requires self-awareness; you must *"learn to live in what alone is life—the present"* (Maxim 12.3).

Just like a hunter stalking prey, you must learn to track your own thoughts and not let them run wild. Take a breath, steady your aim, and weigh the evidence before making a deliberate decision.

"It is told that one speaks ill of you. This is what is told, not that you are hurt by it."

Meditations, 8.49

Impressions are fleeting and swift, they tempt us to act without thinking. So it all comes down to not allowing oneself be swept away by the current of initial impressions.

Haste leads to regret.

"Thus, always stay with first impressions; add nothing to them of your own, and nothing will happen to you."
Meditations, 8.49

The key is to behold the world with open eyes, unclouded by fear or desire. To peer deep into the heart of things, *"stripping the facts bare, beholding their worthlessness, and so removing their gloried regalia" (Maxim 6.13)*, and then to weigh the worth of your actions to determine if you can do something about it or not.

Given that you can choose the meaning to attach to any situation, you can always choose one that carries you forward, a more empowering one. *"If you welcome the impediment and smartly adapt what was granted, an alternative straightway will take its place" (Maxim 8.32).*

Instead of a "lost lover in despair" dwelling on the pain of a relationship gone wrong, you could choose "Great, now I am free to find a better match."

Your chosen meaning does not change the circumstances. Your ex will not come back because you've chosen not to fall into despair. Yet, how you proceed—depressed or driven—depends exclusively upon the interpretation that you choose.

Facts are like stones: hard and cold, empty of meaning, indifferent to our hearts. They simply are. You've split. It's just a fact. No meaning to be gleaned, no moral to be grasped. But we are neither stones nor facts. We are the ones who give meaning to the world. So choose your meaning wisely. Instead of using a learned, default interpretation, which may or may not serve you well, it's better to pick up one that suits you best.

So why not choose to rise above despair and approach the future with determination and hope? Just don't sweat over the things you can't control. The reason is obvious—you cannot change them anyway. You can only control your reaction. So choose the smartest one (or lack thereof) and move on. Ask yourself if there is anything you can do to make things better. If yes, then do it, and do it well. Regard of nothing *(Maxim 6.2, "Just do the right")*.

If you can't, do something else. Work on your goals: learn new skills, do workouts, whatever. But always know what you're doing in life. At any action, ask yourself, *"What does it hold for me?" Shall I repent of it? A little while, and I am dead, all these things are gone (Maxim 8.2).*

Your answer should be crystal clear. Even if it's "I'm just chilling", it's okay. Embrace it. Life is yours to navigate. Follow your own compass, as long as it doesn't steer you towards causing harm to others. You have the freedom to chart your own course, as long as you know your end.

Just stop the musings on the hidden truth. See the world as it is, unadorned and unashamed. When words are spoken, listen well and discern fact from fiction. Then choose the meaning that gives you the most benefit and drive. That brings you strength and joy and poise, and drives you on a worthy course.

In fine, do not extrapolate.
Choose your own meaning.
And act.

Maxim

8.50

"Is the cucumber bitter? Toss it out!

Are there briars in the path? Then turn aside! That's all you need; don't add, "But why are those in the world at all?"

Marcus Aurelius. Meditations 8.50

Each day, we bleed energy denying what's true, resisting facts, lamenting our fate, and worrying over what's yet to come. Angry, sad, and disappointed because reality does not comply...

First of all, never forget: *"Not being angry, but being kind and gentle is manly, as well as manlike"* (Maxim 11.18).

Then, if you *"ever behold as all things come to be through change"* (Maxim 4.36), you should be surprised by nothing, least of all by the things that happen in the natural order of things. Things come and go, like waves crashing against the shore. All is fleeting, nothing stays.

When anger or frustration rises within you, do not blame the world but rather look inward at the expectations you've placed upon it.

Our self-centered expectations, so selfish and grand, like stars, implode into black holes of anger. But align them with reality's tilt, and frustration becomes but a mere flicker. The Universe won't bend to your will, but contentment is still within you.

In the face of life's unyielding waves, we naturally want to ask "why?" as if this question holds the key to our understanding. But "why" is useless. The ocean does not answer the queries. It simply is. To question fate is to waste your soul.

Embrace the vastness of existence. Embrace the chaos. See the world as it is, don't demand it sway. *"Strip the facts bare, behold their worthlessness, and so remove their gloried regalia"* (Maxim 6.13).

Your judgment's a fragile thing, upheld by validating what is laid before you and by never aiming to taint it with the colors of your own projections, *"telling yourself no more than what's declared by the first impressions"* (Maxim, 8.49).

All you can aim for is to acknowledge that what comes to you is not a shock but an occurrence of what you've known, thus finding peace with the present as it is, in all its beauty and fragility, *"loving what happens only, woven in the web. For no fits better, ever"* (Maxim 7.57).

This cucumber is bitter?

Nature ordained it such. A bitter reminder of what we cannot touch.

This love has come undone?

It is the way of things—to shift and change, in the never-ending dance.

Go around these.

Cast aside the bitter fruit. Seek sweeter things to taste.

Find a new love that will not slip away, and hold it tight.

Fired from your job?

The world is wide. Another awaits, one that is meant for you.

Do not question "why?"; accept "what is," and redirect your focus.

Albert Camus, with his absurdity philosophy, states that the human mind is a machine for cause and effect, always seeking reason in the chaos of the Universe. But the Universe, according to Camus, is nothing but a jumbled mess of absurdity, with events occurring without reason or purpose. And so, we are left confused and bewildered, searching for meaning in a meaningless world. But, Camus says, we must push on, setting goals and forging our own path, ignoring the nonsensical grand scheme. We must create meaning in the meaningless and find our own reasons to keep moving forward. For in the end, it is not the Universe that defines us, but the goals we set for ourselves, in spite of the absurdity of it all.

Whether the Universe is absurd or not, is perhaps debatable. But complaining is absurd, for sure. It's simply not a rational way to navigate the world, a waste of breath, and a betrayal of the present moment.

"Fret not at things external. For they care nothing"
Maxim 7.38.

A lot of things in life are minor and not worth sweating over. Remember, minor things are for minor minds. Do not let them linger, clogging the gears of your mind. Do not demean yourself. You simply do not have the luxury of complaining. You don't have time to waste. It will keep you from doing anything of value.

"Do not waste what is left of life worrying about others—save only when your goal is the common good."
Maxim 3.4

Your focus should be on the way *you* move, not the way they do. Marcus Aurelius echoes it like a mantra. *Don't whine. Look within, not without.*

"When got upset by other's failings, pass on at once to dwell on how you do fail alike."
Maxim 10.30

It's easy to complain, to let the words spill out like blood from a cut. But Stoic grace is not in words, it is in action. True strength is facing the world head on. Not griping about what you can't change but tackling the problems within your grasp.

When you discern between what is within your power and grasp and what is not, you realize that it is solely the choices you make, the words you utter, and the thoughts you entertain.

So don't complain. Take action—to shape the reality, if you can do it, or to change the way you see it, if you can't.

Complaints are a curse, a prison cell. Don't just sit and stew. Take a step and ask, "What can I do?" Marcus just tossed the cucumber, but you, you can make something of it.

Your power is action, not lamentation. Actions, not words, will set you free.

It's your journey. Take a charge. Devise a plan. Make moves. Be a man.

"You must arrange life act by act, and be content if each act, as it may, fulfills its end; no one can hinder you in that."
Maxim 8.32

To journey is to know obstacles will come, as sure as sunrise. So carry within you the wisdom of turning stumbling blocks into stepping stones. As Marcus suggests.

"If you welcome the impediment and smartly adapt what was granted, an alternative straightway will take its place that fits the course you are arranging, act by act."
Maxim 8.32

Let the complaining fall away like so much ash, and face the obstacles along the way head-on, swift and sure. Remember Oscar Wilde:
"A pessimist is somebody who complains about the noise when opportunity knocks."

Anyway, either you rise to meet the challenge life throws your way, or... or you crumble. Why waste breath on whining?

"All that happens, happens as given by nature to bear, or not. If it's by nature to bear, then bear it. Do not complain. But if it is by nature not to bear—don't complain; you are to end as strength consumed."
Meditations, 10.3

Complaining also steals the present.

In fact, we always complain, even if it's just internally. We scream silently, whining to ourselves about tomorrow and yesterday. We tell ourselves that happiness lies in a future "when." We fuss over anxiety and wallow in past depression. But it's all an illusion. When that future arrives, the happiness will be fleeting, and we'll realize that nothing external can give us what we already possess in this moment, forever.

This truth is simple. If you're aching, it means you are living the past. If you're fretting, you are living the future. If you are in bliss and lightness, you live in the present, now.

"Learn to live in what alone is life—the present, then you'll be able, for life's remainder and till death, to live at peace with genius within, tranquil, ingenuous, serene."
Maxim 12.3

Marcus' bitter cucumber allegory is a simple one, almost a joke in its own way. A lesson in laughter. In the face of hardship, to either complain or to rise, with grace and reason. When the glass shatters, pause and smile, put on your shoes, and sweep up the pieces, with care and poise.

It's always in your power to see the situation for what it is *(Maxim 6.13, "Strip the facts bare"; Maxim 8.49, "Tell yourself no more than what's declared by the first impressions")*, to understand what can and can't be done, and act with purpose *(Maxim 8.32, "Arrange life act by act")*. No lamentations, no self-pity.

Life at times is really bitter. But most times, it's just a bitter cucumber.

Confucius would say:

A matchstick's strike,
a candle light—
better than cursing the endless night.

The Martyrdom of Saint Livinus, Cornelis van Caukercken,
after Peter Paul Rubens, 1657
Saint Livinus is attacked by pagans who tear his tongue out of his mouth with tongs. A man with the pliers gives Livinus' tongue to his dog. Angels bring from heaven a laurel wreath and a martyr's branch for Livinus. Two bigger angels drive off the villains with thunderbolts.

*Prayer (Precatio) Overcomes Hypocrisy (Fucus), Johann Sadeler (I),
after Maerten de Vos, 1579*
The personification Prayer (Precatio), with her hands folded. Prayer tramples on the personification of Hypocrisy (Fucus), a monk with a rosary and a dagger.

Maxim 9.40

"And who told you the gods aid us even not in what is up to us? Begin to pray for that, and you will see. One prays, "How may I lie with her?"
You, "How may I have no longing for it?" Another, 'How may I be rid of that man?" You pray, "How may I have no wish to be ridden?"
The other, "May I not lose my little child!" You, "May I not be terrified to lose it?"
In sum,

TURN YOUR PRAYERS THIS WAY, SEE WHAT COMES."

Marcus Aurelius. Meditations 9.40

According to a legend, once, in the heat of a battle, Marcus' men were trapped, parched, and dying under the sun's merciless gaze, surrounded by the enemy on all sides. But in that moment, Marcus prayed, and the heavens answered. The sky opened up like a wound, pouring down rain like blood on the battlefield. a miracle rain that legionaries caught in their helmets and drank as they fought their way to freedom. They fought with renewed strength, and the Quadi fell before them like blades of grass in a storm.

This was *"The Rain Miracle"* of Marcus Aurelius.

Meditations 9.40 sees Marcus delve deeper into the subtlety and philosophy of prayer. He wonders why those who trust in divine power would ask for the world's gifts instead of inner peace and flourishing. With a hint of irony, he asks, "Why do we beg for rain, for blessings, when what we truly need is the strength to bear the drought, to neither crave nor fear the weather?" This form of prayer is not a plea for salvation but a way to voice the philosophies we choose ourselves. It is an acknowledgement that within us lies the ability to shape our reality, to shift the tides of our own lives.

Like affirmations, it empowers the soul, whether you're a believer or not. You may be an agnostic or an atheist, but you'll find yourself "praying" at times. There will come a moment when you find yourself reaching out in a plea for something beyond yourself. Even if it's not an actual prayer, just a yearning cast outward, a muttered "Please," "Let me…"

You are likely get so caught up in what you do not want that you are missing what it is you truly do. You focus on the outside. But the world is beyond your control, only your thoughts and actions are in your hands. Longing for external things leads to pain. Find instead the inner fortitude and wisdom to navigate the challenges with grace, rather than yearning for a change in circumstances.

Try a shift in your perspective. Focus on what's within your power. Tend to yourself. Take charge of your health to avoid a serious illness, and there is no need to desperately pray for a miracle later.

Start praying differently.

No longer pleading to the gods above, no longer seeking for a transformation from without. Turn inward, seek wisdom and grace—the fortitude to face the challenges of life.

Don't pray for what you wish to be. Instead, ask how you can find serenity in any outcome, how you can find a way to make the most of it.

Pray to let go of fear, let go of desire, let go of envy—pray to release the chains of all that holds you back.

Learn to find blissful joy within by being a man of integrity and lending a hand to those in need. This inner peace, this quiet joy, it's yours to command, and it'll bring greater happiness to you and those around you.

Do not pray for the world to be kinder. Instead, pray for the courage to confront it with love in your heart. Stop praying for others to improve, and pray for yourself to become better equipped to navigate the complexities of human nature.

Let go of the fixation on the faults of others and the desire for them to alter their ways. This will only bring sadness and sorrow. When reality fails to align with your wishes, it is vital to examine your own failings and ponder ways in which you could have acted with greater grace *(Maxim 10.30, "Dwell on how you fail too")*.

Do not demand change from others. Pray for yourself to be the change, to be the role model, a gentle influence that will inspire growth in all those you touch.

Work hard on shifting your perspective.

What do you seek beyond yourself? A want for the external, how can it be internalized? How can you imbue that pursuit with inner fire?

Stop craving fleeting joys. Question yourself: "How can I attain a serene state of mind, unshaken by the external and the transient?"

That is a prayer, a mantra,
A way to face the chaos
With grace,
That is the Stoic way to pray.

Fable of the Rhinoceros and the Elephants, Aegidius Sadeler (II), after Albrecht Dürer, 1608
A rhinoceros in the foreground. Four elephants in the background. The moral of the story teaches that you must remain true to yourself, even in difficult circumstances.

Maxim
10.8

"ASSUMING TITLES, *good, modest, true, wise-, fair-, great-of-heart,* MAKE SURE YOU NEVER BE CALLED BY A NEW NAME"

Marcus Aurelius. Meditations 10.8

This book is a collection of maxims carved by the mind of Marcus Aurelius, Emperor of Rome. A set of rules to abide by in any situation, a default to turn to when faced with the toughest of choices *(Maxim 3.13, "Always have your principles at hand")*.

Now, dear reader, It is time for you to unearth your own brilliance, and in it take pride, for it is time to let your light shine forth, unabashed and unafraid. Define your *Titles For The Self* as Marcus did.

As you ponder upon them, you learn to trace the contours of your being, and in doing so, you find clarity in your actions. These names are like knives, carving out a shape for the actions they imply. A sharpness to the way you move, a precision in how you define yourself.

With a steady focus on your guiding principles, you can quickly apply these titles as blueprints for the actions you take, turning these labels into weapons that cut through the chaos of choice with precision and purpose, your character etched in every strike.

When you stumble, forgive yourself for the failures; the ones who cannot acknowledge their own worth are doomed to fall again. But rise up, my dear, and claim your own titles; do what must be done. In the act of rising, we become who we are meant to be.

"And if you do depart from these titles, return to them quickly."
Meditations, 10.8

Write the titles down. Let them spill onto the page like confessions, each line a plea for a different version of oneself. To have the main ones illuminated and inked onto paper will aid in etching them into memory, carving them into the skin of your thoughts.

Keep it simple, like Marcus did, for even the most powerful man in the world knew that sometimes all it takes is six words to define the empire within:

"Good, Modest, True, Sound-, Open-, High-minded."

Marcus could have been contemplating the engravings on his own coinage while choosing those names. The town of Tyraa in Scythia etched "Virissimus" upon its coins, a nickname bestowed by Hadrian in Marcus's youth. A title meaning *"most true"* in Latin, that seemed to stick, perhaps due to Marcus's reputation as a philosopher and seeker of knowledge.

In ancient Rome, names were like a labyrinth; emperors' names changed like leaves in the autumn. The young Marcus Aurelius carried the name of his father, Marcus Annius Verus, who passed away, leaving a mystery behind, perhaps when Marcus was but four. Later, Marcus was adopted by Antoninus Pius and given the name Marcus Aelius Aurelius Verus. He succeeded Antoninus and became Marcus Aurelius Antoninus, taking the family name and cognomen of his predecessor. Today, he's known simply as Marcus Aurelius, but during his reign, he was typically referred to as Antoninus or by his imperial titles: Imperator, Caesar, and Augustus.

Cassius Dio, historian of Rome, writes that Emperor Hadrian, who once knew the boy Marcus, was drawn to him after the passing of his father, seeing in him a future emperor *"because he was already giving indication of exceptional strength of character."*

Dio adds:

"This led Hadrian to apply to the young man the name Verissimus, thus playing upon the meaning of the Latin word."
Dio Cassius, Roman History, LCL 176

Hadrian, who saw himself as a poet and loved wordplay, elevated Verus, the "true" one, to Verissimus, meaning "truest" or "most true." Perhaps Marcus, in his youth, spoke words of raw truth before the emperor's eyes. However, Verissimus can also mean "most fitting," so it could be a subtle nod to Marcus' fitness as a ruler, chosen by Hadrian to take the throne.

Anyway, the name became forever linked to Marcus as a philosopher, a symbol of his devotion to truth and the depth and sincerity of his philosophical pursuits.

Now, get your names. Gather them like shells washed up on the shore. Each one a testament to the ocean within you.

Being who you yearn to be starts with a definition. How can you become who you long to be if you do not first articulate the shape of your longing?

Define your wants, become your wants. Action follows clarification.

So, plan the lines, then carve them deep.

Titles for the self!
Let them be carved in blood and bone,
Etched in the skin you call your own.

Maxim

10.16

"No more reasoning at all what is a good man. Be the one!"

Marcus Aurelius. Meditations 10.16

This is Marcus, with words as sharp as a sword. Succinct and direct, at his absolute best.

Right and wrong clear. No time to debate what's a good man.
Be it.

P.S.

Tomorrow, perhaps, you'll find a new path to become a better man. But today, do not forget the beauty of being a good man in every way you already know how. Don't use perfection as an excuse for inaction. Whether you've studied philosophy for a day or a lifetime, like Marcus, the realization that perfectionism is a prison that keeps you from making true progress is a moment to start being the best you can this very instant now, even if you don't have all the answers yet.

Stoicism is not a cold set of rules, it is a burning flame within your heart. True understanding comes from living it, not just admiring it from afar. The Stoics saw the value in a philosophy that was clear and simple and focused on the most important practical questions in ethics.

In the ancient world, Diogenes the Cynic and Plato stood in sharp contrast, representing two distinct attitudes towards philosophy after Socrates's passing. Diogenes looked down upon Plato, scornful of Plato's love for the academic world, his obsession with abstract thoughts and long-winded discussions. Diogenes saw it as a diversion from the true path of practical virtue. In return, Plato called Diogenes a mad Socrates. The Cynics, with their disdain for the superficial, rejected the study of Physics, Logic, and all forms of bookishness as mere Sophistry, distractions from the pursuit of practical wisdom. Philosophical debate like that in Plato's Academy is good only if it helps one be wise and virtuous; otherwise, it's a mere trap of vanity and superficiality. Marcus persistently counts his blessings for being fortunate enough to avoid the latter.

"[I'm grateful to the gods...] that I got passion for philosophy without falling in with any sophist, not stucking at a desk in writing treatises, or solving syllogisms, or dwelling upon physical phenomena."

Meditations, 1.17

The Stoics understood the weight of Logic and Physics, but they also knew the danger of becoming lost in them.

Keep virtue as the goal.

Stick to what's right and don't waste time debating ultra-fine details.

"Time is never enough—enough is the action that you take in a given time."

The ocean of words we spill defining a good man could have filled the cups of our own becoming.

"No overdress of thoughts in finest language. Too many words, too many deeds to be avoided."
Maxim 3.5

Epictetus also advised to speak little in general,
"Be mostly silent; or speak merely what is needful, and in few words" (Enchiridion, 33).

And in particular, keep silent on the principles you've learned, *"for you are in great danger of blurting out some undigested thought."* As even sheep *"do not bring grass to their shepherds and show them how much they have eaten."* Instead,
"they digest their fodder and then produce it in the form of wool and milk" (Enchiridion, 46).

Don't speak of principles, but instead reveal the deeds born from digesting them whole.
Deeds show creeds.
Words like smoke, dissolve in air. So don't speak of your passions, your love, your beliefs. Prove them with action.

Words are empty,
Action is key.
Don't talk the talk,
Walk the walk,
And be.

*Fable of the Pig and the Warhorse, Aegidius Sadeler (II),
after Marcus Gheeraerts (I), 1608*
A pig lies in the mud and watches a passing warhorse. The fable describes a conversation between the pig and the warhorse, where the warhorse disapproves of the pig's life. The moral of the story teaches that an honorable existence is better than a lazy and useless life.

'Triumphus Mortis', Matthäus Greuter, 1596

Maxim

10.29

"At each single act, pause and ask, if loss of this makes death a terror?"

Marcus Aurelius. Meditations 10.29

Death is a master of light, casting shadows on the insignificant. Clarifying, distilling, and revealing what is vital and what is waste. Remembering death—*Memento Mori*—forces us to reassess priorities, opens our eyes to the importance of our choices, urging us to act now on what we can shape.

Just stop and think, *"If death came now, would this moment be worth it?"*

Pause. Listen to the inner silence. Ask yourself, *"Would my spirit be at peace with what I'm doing in these last breaths of life?"*

Your life—the moments you live—is all you have *(Maxim 12.3, "Train yourself to live in what alone is life—the present")*. Each moment is a gem, it's only through what you do in this moment now that you can transform your tomorrow.

What are you doing in this moment? Something worth it? Will it lead you towards the life you dream of?

Imagine a fierce dispute with someone you love. If death came, would you want it to be how you're remembered?

Or perhaps you're drowning in earthly treasures—things that weigh you down, things that were supposed to bring happiness and joy but instead leave you empty?

A journey of self-discovery, of eliminating the non-essential, of *"value clarification"* is a journey without end. A constant shedding of what does not matter. A daily stripping down to the core of what does. Eliminate the excess, and rediscover what brings you life. Make it a ritual of every sunrise.

Confronting oneself and asking the hard questions can be a difficult task. The depths of self-discovery are treacherous waters, but the treasures hidden within are worth the risk, and the rewards are infinite.

Like Socrates before you, you delve into your own thoughts, unearth the inconsistencies in your beliefs, and seek the truth through questioning and introspection *(Maxim 8.2, "Of every action ask yourself, "What does it hold for me?")*.

Here are a few more ideas for *"value clarification."*

Take some things you willingly give your time to, like going out with friends or sweating at the gym. Consider their purpose—what drives these choices? What values do they reveal? Why do you do these things? For what does it all burn? Keep asking, keep peeling back the layers, until you reach the core of what you value, what you believe. It may expose what truly matters to you. It may shatter the illusions of some of your endeavors or connections you hold dear.

Apply these ideals to your career. A job you loathe or a company that contradicts your principles erode steadily your inner peace.

Shed toxic ties, establish clear boundaries. Do you live the life you want with all you own? Purge the unneeded, make space for what's vital.

The questions may come like a storm, but don't give up. Be still, breathe deep. Break them down, one by one, and focus on what matters most. Keep mortality close, and let it guide you as you consider the type of person you want to become and the values you wish to hold *(Maxim 7.56, "As one who's dead, whose life till now lived and gone")*.

The filter of *Memento Mori* aids in revealing the true worth of your time, of what is truly important in life. It reminds you to focus on the depth of life, not just its breadth, to cherish life's moments, not its length.

Don't squander life on emptiness. Seek beauty in the present, filling your moments with the things that fill your heart with love, the things that bring you joy and peace. Let every moment reflect your deepest truth, your truest self.

Fill your time with what is true,
The things that make life good to do,
Values and purpose in your sight,
And life is always worth the fight.

Maxim

10.30

"When got upset by other's failings, pass on at once to
DWELL ON HOW YOU DO FAIL ALIKE;
as holding money for a good, or pleasure, or reputation, and the kind."

Marcus Aurelius, Meditations 10.30

The most beautiful thing about Marcus Aurelius was the way he approached self-evaluation with grace, unburdened by pride or the weight of his emotions. He was driven by a thirst for understanding, to see the truth in everything he came across. In Meditations, he repeatedly ponders his true intent, seeking the origin of his drives and emotions. The fact that he was an emperor, a man above all men, with no one to judge or criticize his actions, only adds to the awe-inspiring nature of his honesty.

Often, we judge others for their misbehavior and imagine ourselves as superior. But this is just an illusion, for we remain ignorant of our own shortcomings. Self-awareness is a painful truth; it's a sharp blade that cuts deep, and we prefer to avoid its sting.

So we persist in our belief that we act rightly, yet sometimes our actions are driven by anxiety, wrath, desperation, or ignorance. This may not excuse us, but comprehending our actions can help us comprehend others, even if they harm us. We all make mistakes, as we are all humans. Thus, before judging others, remember that you too are fallible.

Only when you embrace your vulnerability can you transcend it. And in doing so, you find that the people around you become more relatable and less repelling.

You must confront your emotions and ask yourself why you react in certain ways. Only then can you rise above your shortcomings and strive towards your purpose in life. If this book on Stoicism is any indication, your aim is to better yourself, to become the best version of yourself.

"Let philosophy scrape off your own faults, rather than be a way to rail against the faults of others."
Seneca, Moral Letters, 103

Seneca shows us what philosophy is for. We aim to shed our faults. Philosophy, as the Stoics saw it, was a means to cleanse our own *psyches* (souls). Taking charge of cleaning up your own mess instead of searching for others to blame is a beautiful purification.

Waste not your breath complaining of what others do. Changing them is unlikely, but you can always change yourself—how you respond to their mistakes included.

"It is absurd, not to escape one's own fault, which is fully possible, but trying to escape the others', which is not."
Meditations, 7.71

Escaping your own flaws, defeating them, will change your future. Running from others' flaws will not. The moment you change, everything else follows.

Clean up your own mess. Don't look for others to blame.
It's freeing.

A man is held up a mirror (See for yourself), Johann Theodor de Bry, 1596
Two men sit side by side on the floor. The left one has glasses in his hand. The judge points to a man standing in front of him and holding a mirror up to him. An old woman is standing next to the man with the mirror. Portrayal of the saying 'Besiet uselve', look at your own mistakes before accusing anyone else.

Meekness Overcomes Rage, Pieter Jalhea Furnius, 1550 - 1625
The personification of Fury (Furor) lies on the ground at the feet of Meekness (Clementia). The print has a Latin caption and is part of a series on virtues and vices.

Maxim

11.18

"And let it be at hand in anger that not being angry, but being kind and gentle is manly,

as well as manlike."

Marcus Aurelius, Meditations 11.18

Marcus believes anger to be like a chink in the armor, a vulnerability that shatters at the slightest touch, spilling out in a violent eruption. But sinew, strength, and fortitude do not belong to those who are quick to rage. They are only found in the gentle, the wise, who can keep their peace even in the storm.

Marcus's father was gone before he even had a chance to ask his name, but he was adopted by a noble destined to be Antoninus Pius, Emperor of Rome. Antoninus Pius became Marcus' north star and a role model. Marcus was set to *"See clearly what governs their minds, see what the wise indeed avoid, what aspire" (Maxim 4.38)*.

Meditations detail the traits Marcus admired most in Antoninus Pius and aimed to copy. He began with "gentleness."

Antoninus Pius was

"Nowhere harsh, implacable, or violent, never forcing things, as one says, "past the sweating point."

Meditations, 1.16

Further in Meditations, Marcus delves deeply into Stoic ways of taming anger. The Stoics recognized that there are some emotional reactions that are impossible to avoid in certain situations. Anger is one of the unhealthy passions, like fear or anguish, that a person can overindulge in, exceeding the limits of reason.

The Stoic doesn't fan the flame of anger. Marcus says this shows true weakness. Your initial response may slip from your grasp, but if you are present enough *(Maxim 12.3, "Learn to live in what alone is life—the present")*, then upon *"telling yourself no more than what's declared by the first impressions" (Maxim 8.49)*, you can decide if you carry on or not.

Anger, then, is a kind of judgment or, better said, it clouds judgment, leading to predictable thoughts of retaliation. The insult is perceived as unjust, justifying the anger. But it is only a mask for your own fear and inner pain. Anger is like madness that whispers, "Strike back, they deserve it," painting your perceptions with a desperate need for justice after being hurt. This suffocates your mind, smothering other, more useful trains of thought. But with practice, dedication, and intentional action, you can train yourself to bring forth these positive views when required.

Don't let anger control you. Counter its fire with coolness. Loosen the tension in your face, take a deep breath, let it fill your lungs, and hold it there for a moment. Speak softly, slow down. Your inner peace will soon match your outward appearance. Let the calm seep into every muscle and every thought. Let it weaken the hold of misfortune and grant you the courage and the strength to be productive, composed, and brave, to move forward with grace. Marcus says:

"The freer the mind from passion, the closer the man to strength."
Meditations, 11.18

In general, the world does not conspire against us; it just is. Things just happen, not against us.

"Fret not at things. They care nothing."
Maxim 7.38

Fools we are, letting trifles upset our peace. Marcus advises never to forget the fleeting nature of the world around. *"Ever behold as all things come to be through change" (Maxim 4.36)*. This anger now will be tomorrow's forgotten memory, a soft whisper, lost in the wind.

When anger starts the fire inside of you, try to extinguish it with the cool clarity of objectivity, to describe the situation without emotion or drama, as a detached observer, *"stripping the facts bare, beholding their worthlessness, and so removing their gloried regalia"* (Maxim 6.13).

This will grant you the mercy of time, permitting you to view the situation from a wider perspective. Remember, it is not the circumstance that hurts you, but the thoughts you attach to it.

"Thus, always stay with first impressions, add nothing to them of your own, and nothing will happen to you."
Maxim 8.49

So, with grace, aim to reject the first stirrings of anger and rise against its birth. Choose courage, the embrace of peace, over the storm of anger. Return it with unyielding wisdom, patience, and compassion.

Be brave. Be calm.
The prize, a wise heart, and a kindness
As enduring as the stars.

Maxim 12.3

"Learn to live in what alone is life—the present,

then you'll be able, for life's remainder and till death, to live at peace with genius within, tranquil, ingenuous, serene."

Marcus Aurelius. Meditations 12.3

The present moment seems to be the most obvious, real, and palpable thing there is. And yet it eludes us completely.

All the pain in life comes from that fact.

Every second, our senses are bombarded with stimuli (whether we realize it or not), and a parade of emotions and thoughts pass through our heads.

Each moment represents a whole universe, forgotten next instant when being replaced with an apparently similar one. But we can't let the clock or the cycling calendar make us blind to the fact that each moment of life is a marvel and mystery. It's just a matter of paying attention to this miracle.

The first step for Stoics is self-observation, and noticing what you're doing from moment to moment, being especially conscious of value judgments.

Stoics believe that the good can only reside in the "here and now" as that is the place where our volitions start. Yet everything in life seems to conspire to make the thoughts go stray from where they originated. The mind is like a butterfly that flits from flower to flower. We turn less mindful and more mindless as this happens again and again. Most people then seek happiness in a roundabout manner, by chasing external things they hope to obtain in the future.

For Stoics, the past and the future are "indifferent" because the present moment is the only seat of our volition, where virtue, "good" or "evil" can possibly originate.

Stoics therefore train themselves to pay attention to the present moment, while accepting "indifferent" things with tranquility. They focus on becoming good right now, on cultivating wisdom and justice in actions, which is the direct and only route to flourishing and happiness, or *eudaimonia*.

To sum it up, the most precious thing in the universe is contained within you, right here in the present moment, where the past and the future, the two of eternities, meet.

But you are not living in eternity. Now—is your time to become good. All you have is this moment, *"sparkling like a star in your hand—and melting like a snowflake"*.

Live in the moment, in the breath, in accordance with nature. Be continually mindful that good and evil are in your choices this very moment, not in anything external.

Sanguine temperament, Harmen Jansz Muller, after Maarten van Heemskerck, 1566

Jupiter, with lightning and eagle, and Venus, with arrow and Amor, sit in the sky and represent the planets of the same name. Above them are three signs of the zodiac: Libra, Gemini and Aquarius, which correspond to the element of air. Among them, on earth, the people, who have a sanguine temperament: elegant couples in love, people bathing and dancing in the open air. At the bottom in the margin is a text in Latin about the sanguinici and the associated planets.

Maxim

12.6

"Practice, even when despair of success.

The left hand, idle at other tasks for lack of practice, still holds the bridle stronger than the right, because of practice."

Marcus Aurelius. Meditations 12.6

At first, this Maxim looks like a worn tale of perseverance and triumph, success through steady steps. Still, we need continual reminders of these universal truths. They, like stars, need constant shining. Marcus illuminates them with his clarity, sharp and dazzling as always. And this one is really simple: impossible is just a word.

Let this lesson etch itself into your bones—you hold within you the potential to soar beyond your wildest dreams. When the world whispers, "It's impossible," whisper back with your actions. Take on the challenge. Practice in what seems impossible.

Practice is the master of all things, it makes everything perfectly possible. This is often misquoted as *"practice makes perfect"* ;)

"No one can teach riding so well as a horse."

This seems to be quite a common-sense wisdom, but Marcus' focus is on practice, perseverance.

Knight, Death and Devil, anonymous, after Albrecht Dürer, after 1513
A knight on horseback, riding along a rocky path with his dog beside him, encounters Death, also on horseback, and the devil on his way. In the background is a city.

The key is:

"Nothing is impossible until you quit."

Or

"Nothing is impossible to industry."

Brick by boring brick, the wall is being built *(Maxim 8.32, "Arrange life act by act")*. Thus, in stages, the impossible becomes possible.

It's easy to turn away from what's hard when it seems impossible at first. To attain greatness in any difficult endeavor, one must persevere—practice despite negativity, transcending skepticism, transcending pain. Markus offers a simple truth: Is it possible at all? Then do it.

"If something's hard for you, do not assume that for a man it is impossible, but whatever is humanly possible, consider also attainable for you."
Meditations, 6.19

Think of the things once thought impossible by humankind: journeying across the endless ocean, flying with the birds, touching the moon, or instant messaging across the continents. All these awe-inspiring achievements started with a single flicker in the dearing imagination. Yet, there are also the delicate, personal battles, each a mountain to be climbed, no less challenging and overwhelming: financial security, good health, learning a new language, or the freedom to articulate in public.

Those goals may seem like distant stars, but they are not unattainable. With practice and determination, you can reach for them and touch them with your own hands. Moreover, the more you practice the "unattainable," the more you bleed for it, the more success you find.

Embark on your tough journeys. Begin to do hard things.

Believe that you can birth the dream once deemed unfeasible. Fear of failure—the only thing that keeps the dream out of reach. Remember, what you fear is often not reality but a mirage created by your mind's eye. Beat fear with preparation and reason *(Maxim 6.31, "Sober yourself: recall your senses")*.

After all, "impossible" is not a fact. It's an opinion. Nothing is impossible unless you think it is *(Maxim 8.49, "Do not draw inferences in excess")*.

Observing things from a different perspective could allow to remove the "impossible" regalia, things are sometimes wrapped in *(Maxim 6.13, "Strip the facts bare")*. Observing is seeing what is. Perceiving sees beyond; more than what meets the eye (Marcus speaks of exclusive wine as nothing more than old, fermented grapes, stripped of glamor). To observe without perceiving takes both patience and determination. When faced with an obstacle that makes a task seem "impossible," shift the lens, release yourself as the perceiver, and enter into the proper context. Delve deep, allow your thoughts to breathe, and discover the unexpected solutions to what at first seemed impossible.

To paraphrase Winston Churchill: "The Stoic sees the *invisible*, feels the *intangible*, and achieves the *impossible*."

Ways exist for all things. With enough will, there's always a means. To call something "impossible" is to simply avoid the challenge.

"Impossible is not a declaration. It's a dare. Impossible is potential. Impossible is temporary. Impossible is nothing."
Muhammad Ali

Get it in your brain:
Only nothing is impossible.

Honor, Philips Galle, c. 1585 - c. 1590
A standing man with a long cloak and a laurel wreath around his head. In his right hand he has a text scroll, in the left hand a stick against which a shield rests. A naked woman is depicted on the shield.

MAXIM 12.17

"IF NOT RIGHT, DO IT NOT; IF NOT TRUE, SAY IT NOT.
Stand firm on yours."

Marcus Aurelius, Meditations 12.17

The pages of Meditations are stained with Marcus' devotion to truth, referenced repeatedly. He even goes so far as to adopt the names of various virtues, including *"True."* The word he used was *alethes*, as he wrote in Attic Greek, not Latin. Marcus, beyond doubt, was aware that this could be seen as a translation of his family name, Verus, meaning "true" (*Maxim 10.8, "Take your Titles"*).

Maxim 12.17 holds the keys to living with honesty and integrity. Honesty and integrity, though often spoken in the same breath, are not one and the same. A man may be honest yet lack integrity. But integrity cannot exist without honesty at its core.

Honesty is being truthful in words. Transforming truthful words into reality is integrity. To speak the truth is not enough, truth must be lived. Action is the pulse of integrity, pulsing with honesty's beat.

Thus, integrity would be: *"doing as promised, when promised, as stated."*

It would mean being honest with yourself, first of all. *"To thine own self be true, and it must follow, as the night the day, thou canst not then be false to any man,"* as William Shakespeare put it.

Stay true to yourself. Think only what aligns with your beliefs,
"For the soul takes the dye from its thoughts."
Meditations, 5.16

Who you are is your decision. You are a mosaic of every choice you make, every thought you think, every action you take on each day. Your integrity decides your fate. *"It is the light that guides your way,"* as Plato put it.

In 156 CE, the Christian apologist Justin Martyr penned a letter called "The First Apology," where Marcus Aurelius was referred to as "Verissimus the Philosopher." So, it was not just a nickname from his childhood given by Hadrian. The title clung to Marcus, continuing to echo through his life as Caesar. He was still Verissimus, "The Truest."

Marcus claimed his childhood nickname, making it a symbol of his life's philosophy. In the chaos of Rome and on the battlefields of Central Europe, he persevered with steadfastness and honor, earning the most prestigious title of all—a man of integrity.

Keep your words and life together,

"and if you hold to this, with nothing to await or fear, being content to act by nature in the present and with the truth of times heroic in every your word and speech, then you will lead a good life. No one is able to prevent this."
Meditations, 3.12

Truth, always say it,
Words, less than deeds,
Practice what you preach,
Integrity lost, honor too,
Recovering, beyond reach.

Maxim

12.20

"First, to do nothing at random, or without a purpose.
Second, to DO NOTHING BUT LEADING TO SOME SOCIAL END."

Marcus Aurelius. Meditations 12.20

The first sentence here is nothing else but *Maxim 7.29, "Stop acting like a puppet":*

Remove impulses from your life;

Make your actions purposeful; never waste your energy on nonsense;

Have a goal.

But what goal?—Act according to the common good.

Why?—No noble aim exists beyond that. No other goal brings happiness. Aristotle claimed happiness to be life's meaning, purpose, and end.

The question is: "How to reach that end?"

We all "hunt happiness." But in fact, the majority waste their days chasing a mirage made of the bitter fruit of pleasure-seeking, possessions, and vanity—a mad, self-destructive delusion stemming from a broken world. So they fill their homes with things they don't need, they kiss lips they don't love, they work hard to get approved by people they don't even like.

Why all this, they do not even ask. They follow the flicker of the closest joy, it burns bright for a while in front of them, then scatters with false sparks. They understand they have been fooled, but they already see a new light, and they go for it. They hope this time things will change. But it, too, fades. An endless cycle.

As the day ends and they lay in bed, alone or beside a partner, wondering what's next in this eternal quest, the feeling comes: "*We are alone, alone in this cold, soulless, indifferent, impersonal, and immensely vast Universe, in which our cry is unheard, our role is absent, our worth is but zero, our loneliness is limitless.*"

Thus, lonely beings wander in the mazes of false meanings, and then death catches up with them. Sad story.

But.

The truth is, we're one. All part of a single entity, cells in a single, nearly immortal organism known as the social body of rational creatures. This knowledge, when you really come to it, destroys all doubts and despair. And then the understanding comes that happiness can't be a goal in itself. It's not something you have to "attain." Then, with time running out, what should we aim for? Marcus Aurelius makes it plain in his Meditations, saying repeatedly that as social rational beings, we are meant to support our fellow man.

The goal of life is not happiness or joy. It's *Usefulness.*

Happiness is a result of your virtue and worthiness, a measure of the good done for the good of all.

The wise men of the East believed that if you look deep into yourself, you will realize that there is nothing there. Then why would you care for nothing? And if you look around, there is nothing real anywhere else either. Only choices are there that you fill the void with. And when you do good and feel happy for another, you fill that emptiness with love. Whose love is it then if there is no one anywhere anyway? The emptiness just doesn't care, wise men say. But if you seek the meaning and purpose of your life, you'll never find a better one.

Marcus Aurelius painted other metaphors.

He imagined all rational beings as limbs or organs of the same pulsing, living body, as coordinating pieces of a single organism. To Marcus, the rational in every being was connected, like the limbs of a single entity, meant to function as one, designed to act in unison towards a shared purpose.

Mirror of the Community, Theodore Galle, 1610
A young man and an old man look together in a large mirror and see their reflection. In the background on the left, Pythagoras shows two youngsters a mirror. In the background right, Socrates shows his students a mirror. The print has a Latin caption: 'Naturam iuvat artis opus; specvloque magistro, Praecipit arcans emaculare notas'. At the bottom a legend of the scenes numbered alphabetically in the image.

"For we have come into being for co-operation, as have the feet, the hands, the eyelids, the rows of teeth, upper and lower" *(Maxim 2.1, "Say to yourself in the early morning: Today I shall encounter a busybody, an ingrate, a bully, a liar, a schemer, a self-seeker").*

The Stoics see us all as sharing a single, fragile vessel (now we would say a spaceship) called planet Earth. We are all in the same boat. Our survival is linked, our happiness dependent on each other working together to keep the ship afloat. What brings the greatest joy to a rational and social being is to be of use in those efforts. To build something, anything, for oneself or others, that is the path to joy and satisfaction, true happiness, a life well-lived.

Ralph Waldo Emerson put it this way: *"The purpose of life is not to be happy. It is to be useful, to be honorable, to be compassionate, to have it make some difference that you have lived and lived well."*

It's as simple as this: *What have you offered to this world? What mark have you left? Did your heartbeat serve a greater good?*

Being useful is a mindset, a state of being, arising from a decision to make a difference. Tomorrow, right in the morning, ask yourself: What is your offer to this world?

If there is silence, start to give.

It could be painting, creating, helping those in need, or any other expression of your soul, any act of kindness that feels true. Don't hold back, don't overthink it. Just do something, anything, to leave your mark.

You don't have to transform the world, just leave it a little bit more beautiful than it was when you came. A life well-lived is built through tiny acts of grace and love each day. A life that makes a difference to lives of others, a life that echoes for long.

"You must arrange life act by act, and be content, if each act, as it may, fulfills its end; no one can hinder you in that" *(Maxim 8.32, "Arrange life act by act")*

And as you navigate your way, remember these words of Marcus:

"Life is short, one fruit of earthly living is righteous attitude and action for the common good."

Meditations, 6.30

Maxim

12.1

"And fear not that life will ever stop, but that you never start off living in accord with nature."

Marcus Aurelius. Meditations 12.1

Earthly days enchant and fascinate us. We get lost in the spell, forget what they are turned to with their transcendence... We forget that this is it—one shot. No do-overs. Our time on this stunning blue rock adrift in a Universe both infinitely cold and indifferent is just the only chance we'll ever get.

Living towards death...

We catch the first word, and then the moment comes when fate assertively proclaims the last. And there's only one thing left for a human being who's reached the pinnacle of their spirit: to transform being, turned to death, into death, turned towards being.

This is not seeing the death as "the end," it's the continuous awareness and the reminder that your body and mind are always in a state of transformation and becoming,

"ever beholding as all things come to be through change" (Maxim 4.36).

This is seeing death in its raw truth—as a part of the cycle of life, unapologetically natural. And loving it, *"loving what happens only, woven in the web"* (Maxim 7.57). Amor fati.

This is remembering death, "*Memento mori*"—"*thinking of yourself as dead*" (Maxim 7.56).

Because only by staring death in the face can we find meaning and appreciation in life. Thus we accept death as part of existence, expect it, and remember to live, finding gratitude in every breath we take, embracing life with all its joys and sorrows, "*learning to live in what alone is life—the present*" (Maxim 12.3).

Why is remembering to live crucial ?

At some point, you will die. It's a fact. An inevitable truth.

But have you truly felt alive?

Have you ever lived? Have you burned with a fire or simply been, taking up space, taking up air, and giving off but carbon dioxide? Consider what else you could grasp, what more could you add to your journey? What more of the beautiful things could you do? How could you live with more fervor, more drive?

Have you ever stopped to reflect on those things? Whatever the answers, commit to *MAXIM 4.17*:

"Not to live as if you had ten thousand years before you. The common due impends; WHILE YOU LIVE, WHILE YOU MAY, BECOME GOOD."
Meditations, 4.17

A "good life," the Stoics say, is marked by virtues such as justice, wisdom, self-control, tranquility, kindness, patience, and a courageous heart in the face of all adversity. And death is not an adversity. It's not a problem at all. Death is just death—not good, not bad. It is neither white nor black, neither blessing nor curse. It just does not bow to our will. It's something we cannot control. The real problem is how we judge things we can't control. Instead of fretting over things you can't change, aim to live a good life—a life that invites death to embrace you with grace.

Death is certain. It's a wave, always rolling towards you. The challenge is to ride it with poise, and this can only be achieved by anchoring yourself in a life that sparkles, shaped by the choices you make. It's the decisions you dare to make, the way you burn bright that separate a good life from one that fades into obscurity.

Goodness, Crispijn van de Passe (I), 1589 - 1611
The personification of Goodness stands in a landscape with a burning heart in hand. At her feet stands the pelican, which pierces its breast to feed its young with its blood.

Virtue or Bravery, Philips Galle, c. 1585 - c. 1590
A standing helmeted woman, holding a lance in the left hand and a sword without a point in the right hand. She tramples Satan with her left foot.

A good life is one rich in meaning and truth, one lived so fiercely, so fully, that it makes death itself bow in reverence.

You can get good from yourself. *"Dig within. Within you is the fount of Good"* (Maxim 7.59). The fount of good, it springs up from within, from taking the reins, from choices we make with purpose and responsibility.

Actions with good intentions bring peace; the only chance for joy and happiness. Just take small steps towards good, whenever possible, whatever the outcome. *"Just do the right. Regard of nothing"* (Maxim 6.2).

Do good because it's right, with no thought of reward.

"Like the vine, which has produced a bunch of grapes and looks for no reward beyond once it has borne its proper fruit."
Meditations, 5.6

Do good for the sake of good itself. It's your nature. It's your job.

"What's your job? Being good."
Meditations, 11.5

That's the simplest job description, ever. But it ain't a cakewalk. Aim for goodness, and you'll find your way. One good act leading to the next, like Marcus Aurelius, the mightiest man in the world in his reign, reminded himself: *"You must arrange life act by act and be content if each act, as it may, fulfills its end; no one can hinder you in that"* (Maxim 8.32).

So, take a pause. Think on the things that bring you joy and make you feel alive, part of this world. What are they?

Is it true that you *"do nothing at random... nothing but leading to some social end"* (Maxim 12.20)?

Not really?

Then when will you be bold and act?

Fearing not death but fearing not having lived. Becoming good.

When will you break the chains,
Of fear and refrain,

From living fully, with pride,
Beauty and virtue, side by side?

Being brave, being good, being live.

The End

Acknowledgments

My personal journey, the fruit of which is this book, would be impossible without generous souls sharing their wisdom at:

donaldrobertson.name
howtobeastoic.wordpress.com
medium.com/stoicism-philosophy-as-a-way-of-life
mindandpractice.com
modernstoicism.com
orionphilosophy.com
stoicself.com
thestoicgym.com
whatisstoicism.com
thestoicsage.com

SPECIAL BONUS
Please Take This Book For Free!

"Memento Marcus: Essential Meditations of Marcus Aurelius"

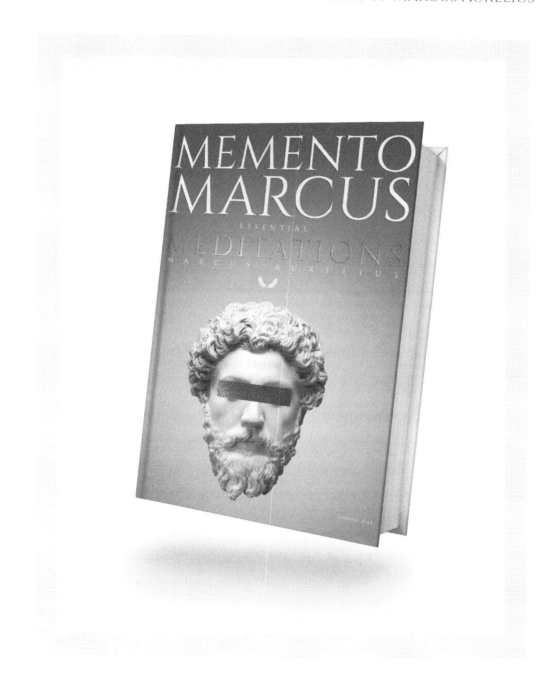

What if you could use the Meditations of Marcus Aurelius as the structured meditation manual for achieving inner strength that only tranquil and mature mind can offer?

"Memento Marcus: Essential Meditations of Marcus Aurelius" aspires to sort out main themes of *Marcus Aurelius' Meditations* by uncovering a scheme that underlies the apparently unordered composition of the Emperor's text and seeks to inspire the process of deeper meditation on select and ordered Marcus' maxims by adding illuminative material : *60+ allegorical frescoes of the 17th-century Dutch artists!*

Get FREE Access To Future Releases By Joining ATLAS OF WISDOM Books Club!

SCAN with
Camera To Join

List of Illustrations

Portret van Marcus Aurelius, John Faber (I), after Peter Paul Rubens, after anonymous, 1691 - 1721

Only a sick body is cared for, Philips Galle (attributed to workshop of), after Philips Galle, 1610 - 1676

Allegorical five-headed monster, anonymous, 1575 - 1618

Memento mori, Raphaël Sadeler (I), after Christoph Schwarz, 1687 - 1749

Portrait of Julius Caesar, Aegidius Sadeler (II), after Titian, 1624 - 1650

Young Man with a Skull, Lucas van Leyden, 1517 - 1521

The Virtuous Man, Crispijn van de Passe (I), 1589 - 1611

Pride , anonymous, after Frans Floris (I), 1575

Wrong belief alienates the world from the truth, Dirck Volckertsz. Coornhert, 1575 - 1581

Moderation, Jakob Frey (I), after Domenichino, 1721 - 1725

The Four Elements, Jacob Matham (attributed to), after Hendrick Goltzius, 1588

Ancient busts of Socrates, Paulus Pontius, after Peter Paul Rubens, 1638

Death of Socrates, Jean François Pierre Peyron, 1790

Father Time Takes Life Away, Bernard Picart, after Carlo Maratta, 1693 - 1696

Sentry Sitting on a Stone Block, Adolphe Mouilleron, after Joseph Nicolas Robert-Fleury, 1851 - 1862

Apollo, Pan and a putto blowing a horn, Giorgio Ghisi, after Francesco Primaticcio, 1530 - 1582

Still life of wildlife , C. Le Coq, after Dirk Valkenburg, 1840

The Dream of Reason Produces Monsters, Francisco Goya, 1799

Unity (Concordia), Cornelis Cort, after Frans Floris (I), 1560

Fable of Cupid and Death, Dirk Stoop, 1665

Each thinks his owl to be a falcon, Hendrick Goltzius, after Karel van Mander (I), 1590 - 1594

Allegory of Life, Giorgio Ghisi, after Raphael, 1561

Personification of Joy of Life, Melchior Küsel (I), after Johann Wilhelm Baur, 1682

Don Quixote Destroys a Puppet Show, François de Poilly (II), after Charles-Antoine Coypel, c. 1723

Fable of the Bear and the Bees, Aegidius Sadeler, after Marcus Gheeraerts (I), 1608

Aurora and the Zodiac, Jacob van der Schley, 1725 - 1779

Bird's-eye view map of Jerusalem depicting Christ's Passion, Johann Daniel Herz (I), c. 1735

Vanitas still life with skull, Jan Saenredam, after Abraham Bloemaert, 1575 - 1607

The Inconstancy of Fortune; It is Easy to Dance when Fortune Plays for You, Dirck Volckertsz. after Maarten van Heemskerck, after c. 1560 - before 1590

Socrates Looking at Himself in a Mirror, Bernard Vaillant, after Jusepe de Ribera, 1672

Allegory of Time and Fortune, Monogrammist AC (16th century)

Fable of the Tortoise and the Hare, Aegidius Sadeler (II), 1608 - 1679

Strength (Fortitudo), Jacob Matham (attributed to), after Hendrick Goltzius, 1587 - 1637

Fable of the Young Man and the Swallow, Aegidius Sadeler (II), after Marcus Gheeraerts (I), 1608

The Martyrdom of Saint Livinus, Cornelis van Caukercken, after Peter Paul Rubens, 1657

Prayer (Precatio) Overcomes Hypocrisy (Fucus), Johann Sadeler (I), after Maerten de Vos, 1579

Fable of the Rhinoceros and the Elephants, Aegidius Sadeler (II), after Albrecht Dürer, 1608

Fable of the Pig and the Warhorse, Aegidius Sadeler (II), after Marcus Gheeraerts (I), 1608

'Triumphus Mortis', Matthäus Greuter, 1596

A man is held up a mirror (See for yourself), Johann Theodor de Bry, 1596

Meekness Overcomes Rage, Pieter Jalhea Furnius, 1550 - 1625

Sanguine temperament, Harmen Jansz Muller, after Maarten van Heemskerck, 1566

Knight, Death and Devil, anonymous, after Albrecht Dürer, after 1513

Honor, Philips Galle, c. 1585 - c. 1590

Mirror of the Community, Theodore Galle, 1610

Goodness, Crispijn van de Passe (I), 1589 - 1611

Virtue or Bravery, Philips Galle, c. 1585 - c. 1590

*Should you enjoy this book, please kindly leave a review at amazon.com
It means the world to me!*

*Thank you,
Andrii*

Atlas of Wisdom

Made in the USA
Middletown, DE
24 September 2023

39244694R00116